Photodynamic Therapy in Veterinary Medicine: From Basics to Clinical Practice

Fábio Parra Sellera
Cristiane Lassálvia Nascimento
Martha Simões Ribeiro
Editors

Photodynamic Therapy in Veterinary Medicine: From Basics to Clinical Practice

Springer

Editors
Fábio Parra Sellera
University of São Paulo
Department of Internal Medicine
School of Veterinary Medicine
and Animal Science
Butantã
São Paulo
Brazil

Martha Simões Ribeiro
IPEN-CNEN/SP
Center for Lasers and Applications
Cidade Universitária
São Paulo
São Paulo
Brazil

Cristiane Lassálvia Nascimento
Santos Aquarium
Veterinary Unit of Santos Aquarium
Ponta da Praia
Santos
São Paulo
Brazil

ISBN 978-3-319-45006-3 ISBN 978-3-319-45007-0 (eBook)
DOI 10.1007/978-3-319-45007-0

Library of Congress Control Number: 2016959212

Printed on acid-free paper

This Springer imprint is published by Springer Nature
The registered company is Springer International Publishing AG
The registered company address is: Gewerbestrasse 11, 6330 Cham, Switzerland

We dedicate this book to all, humans and animals, victims of cancer and infectious diseases. We expect that our effort into making this book brings more prosperity and quality to their lives. We also acknowledge our family and friends, the pillar of our achievements.

"We can easily forgive a child who is afraid of the dark; the real tragedy of life is when men are afraid of the light."

Plato

Foreword

The clinical use of the photodynamic effect has, thus far, been utilized mainly in human oncology since the latter part of the last century. However, von Tappeiner and Jesionek's initial findings concerning the use of dyes and light in this field had been pre-empted by their student Raab, who demonstrated the antimicrobial effect several years earlier, in 1900. Consequently, it has taken – and, indeed, is taking – a considerable time to progress "from bench to bedside."

The requirement for the triple combination of photosensitizer, light, and oxygen to provide cytotoxicity, whether in cancer or infectious disease, rather than via solo attack via drug, radiation, or surgery seems to have caused considerable reluctance on the part of clinicians to adopt the photodynamic approach. However, the ease of use, anti-resistance capability, and excellent cosmetic outcomes should be persuasive arguments, and it has been immensely frustrating that clinical uptake has been so slow over the past quarter century.

Given that any therapeutic agent licensed for use in humans must undergo significant animal testing, the data available to support photodynamic veterinary medicine are, by now, extensive, to say the least. In addition, there are persuasive arguments for alternative approaches to treating animals in the same way that there are in human medicine, for instance, increased cancer incidence due to solar radiation and the rise of antimicrobial drug resistance through over- and misuse of hard-won anti-infectives. Indeed, in the latter category, the future basis of animal and human infection control is inextricably linked due to previous practices.

Microbial drug resistance can be seen as the fortuitous expression by the organism of genetic mutation. This occurs when the mutation is expressed as, for example, a molecular change in the structure of an enzyme which, in turn, alters the binding of a previously inhibitory drug molecule. The genetic change therefore causes downstream alteration of drug efficacy, and if this promotes the organism's survival, it is adopted by the fact that the organism survives where others in the population without the change do not.

Such behavior obviously requires drug exposure, so where this is unnecessary – for example, in growth promotion – it should be discouraged. However, there are other scenarios where antimicrobial use is necessary, and it is here that the use of the photodynamic approach can make a difference. Local or topical infection – for example, soft tissue wounds or mastitis – is currently treated with conventional antimicrobials, with the concomitant risk of later resistance development. Due to nonspecific modes of action, treatment with photoantimicrobials would not only circumvent this problem but would also allow the conservation of valuable conventional antimicrobial agents for more serious, i.e., systemic, disease. Furthermore, logically, while the local/topical administration of any drug exposes far less of the host to potentially harmful side effects, this is also true for the microbiome. Systemic administration of antimicrobials might kill the target organisms, but it invariably damages "friendly flora" and, more worryingly, increases the microflora's exposure to xenobiotics, also raising the probability of drug resistance development. Clearly this is dangerous from the point of view of the animal involved, but in agriculture, since there is the possibility of the resistant microbes entering the food chain, this is also hazardous to humans.

It is fair to say that, to date, veterinary medicine has been viewed as being of secondary importance in comparison to that of *Homo sapiens*. Whether this is due to animals – particularly agricultural animals – being taken for granted by the greater part of civilization or to financial inequalities in agricultural versus nonagricultural commerce, the fact remains that, in medicine at least, changes must be made.

The advantages of the photodynamic approach, as noted above, offer a positive way forward in animal healthcare with the added bonus that there is no downside in terms of communicated drug resistance seen over time with conventional antimicrobials. Clearly photosensitizers cannot act as a complete replacement for conventionals, either in the field of oncology or infection control; the important point is that they should be employed where possible in these areas, either as a direct substitute for or in adjuvant therapy with standard chemotherapeutics. As the following authors demonstrate, the photodynamic approach can be an amazingly powerful therapeutic tool; utilizing this power for the benefit of animals (and humans for that matter) is usually out of their hands.

Lea, Lancashire, UK Mark Wainwright
May 2016

Preface

Photodynamic therapy (PDT) is an emerging light-based technology that uses light in combination with a photosensitizer (PS) to treat localized tumors and infections. Historically, PDT has had a paramount role in cancer therapy and is also broadly used to treat age-related macular degeneration. For cancer, the highest benefits of PDT comprise the avoidance of systemic treatment, the selectivity for malignant cells, and the repeatability. Regarding antimicrobial PDT, main advantages are the wide spectrum of action (bacteria, yeasts, viruses, and parasitic protozoa), the outcome is independent of the antibiotic resistance pattern of the microbial strain, the wide-ranging decrease in pathogens with minimal damage to the host tissue, the absence of selection of "photoresistant" strains after multiple treatments, and the lack of mutagenicity. In addition, PDT is a low-cost and minimally invasive localized therapy.

The fundamental requirements justifying to move PDT toward the clinical field are satisfactory until now. Thus, it appears very appropriate to write a book, which covers the state of the art as regards the fundamental mechanistic aspects of PDT, as well as the diseases that can be most favorably addressed by this therapeutic modality. This can be of great help to both basic investigators in order to orientate their activities toward the solution of problems emerging from in vivo studies and the clinical veterinarians to better understand the scope, as much as the potential of the technique, and to identify suitable avenues leading to an optimization of the therapeutic protocols.

Our intention was to compose a book that is of easy comprehension, pleasant to read, and interesting to researchers and clinicians. The first seven chapters encompass the history and principles of PDT, including its mechanisms of action on molecular and cellular levels, systemic effects, and a multimodality dosimetry (i.e., the three key components for a successful PDT). Chapter 8 highlights the main aspects to be considered by veterinarians to move PDT for clinical practice. Chapters 9 and 10 report the use of cancer PDT in basic and clinical studies. In Chaps. 11 and 12, we introduce the antimicrobial PDT. Chapter 13 covers the use of PDT in other practices, and the book finishes with future perspectives for the use of PDT in veterinary medicine.

We hope that the reading of these pages may provide a wide and deep insight of the potential of the PDT, as much as may inspire veterinarians to establish its safe and efficient use in veterinary medicine.

Sao Paulo, Brazil

Fábio Parra Sellera
Cristiane Lassálvia Nascimento
Martha Simões Ribeiro

Acknowledgments

The editors express their gratitude to all those who directly or indirectly collaborated in making this book, mainly to our family and friends for their dedication and support. The editors gratefully acknowledge financial support from the following Brazilian research funding agencies: CAPES, CNPq, and FAPESP.

Contents

Contributors

Julia Buchholz Animal Oncology and Imaging Center, Huenenberg, Switzerland

Michael Richard Hamblin Wellman Center for Photomedicine, Massachusetts General Hospital, Boston, MA, USA

Department of Dermatology, Harvard Medical School, Boston, MA, USA

Harvard-MIT Division of Health Sciences and Technology, Cambridge, MA, USA

Cristiane Lassálvia Nascimento Veterinary Unit of Santos Aquarium, Santos Aquarium, Ponta da Praia, Santos, SP, Brazil

Fabio Celidonio Pogliani Department of Internal Medicine, School of Veterinary Medicine and Animal Science, University of São Paulo, Butantã, São Paulo, SP, Brazil

Martha Simões Ribeiro Center for Lasers and Applications, Nuclear, and Energy Research Institute, National Commission for Nuclear Energy, São Paulo, SP, Brazil

Caetano Padial Sabino Department of Microbiology, Institute for Biomedical Sciences, University of São Paulo, São Paulo, SP, Brazil

Department of Clinical Analysis, School of Pharmaceutical Sciences, University of São Paulo, São Paulo, SP, Brazil

Center for Lasers and Applications, Nuclear and Energy Research Institute, National Commission for Nuclear Energy, São Paulo, SP, Brazil

Department of Medical Biophysics, Princess Margaret Cancer Institute/University of Toronto, Toronto, Canada

Fábio Parra Sellera Department of Internal Medicine, School of Veterinary Medicine and Animal Science, University of São Paulo, Butantã, São Paulo, SP, Brazil

Chapter 1
History of PDT

Fábio Parra Sellera, Caetano Padial Sabino, and Michael Richard Hamblin

Abstract This chapter presents the brightest historical milestones behind the development of photodynamic therapy (PDT). We initially present how photodynamic reactions were first observed by scientists from three different countries in the beginning of the twentieth century. Oskar Raab, from Germany, observed by accident that protozoan cells stained with fluorescent dyes were killed upon illumination, while Prime, in France, reported that human subjects who ingested also fluorescent dyes for an experimental treatment of neurological diseases developed severe erythema after short exposure to sunlight. Niels Finsen, from Denmark, was awarded with the third Nobel Prize of Medicine in the history for the development of light-based treatments for skin infections. Following, we describe how PDT slowly evolved until the 1960–1970s when new generations of less toxic photosensitizers were developed for diagnosis and treatment of solid tumors. Only then PDT

F.P. Sellera (✉)
Department of Internal Medicine, School of Veterinary Medicine and Animal Science, University of São Paulo, Av. Prof. Orlando Marquês Paiva, 87, Butanta, SP 13635-900, Brazil
e-mail: fsellera@usp.br

C.P. Sabino
Department of Microbiology, Institute for Biomedical Sciences, University of São Paulo, Av. Lineu Prestes 1347, Cidade Universitária, Sao Paulo 05508-000, SP, Brazil

Department of Clinical Analysis, School of Pharmaceutical Sciences, University of São Paulo, Sao Paulo, SP, Brazil

Center for Lasers and Applications, Nuclear and Energy Research Institute, National Commission for Nuclear Energy, Sao Paulo, SP, Brazil

Department of Medical Biophysics, Princess Margaret Cancer Institute/University of Toronto, Toronto, Canada
e-mail: caetanosabino@gmail.com

M.R. Hamblin
Wellman Center for Photomedicine, Massachusetts General Hospital, 50 Blossom Street, Bartlett Hall, Room 414, Boston, MA, USA

Department of Dermatology, Harvard Medical School, Boston, MA, USA

Harvard-MIT Division of Health Sciences and Technology, Cambridge, MA, USA
e-mail: hamblin@helix.mgh.harvard.edu

F.P. Sellera et al. (eds.), *Photodynamic Therapy in Veterinary Medicine: From Basics to Clinical Practice*, DOI 10.1007/978-3-319-45007-0_1

really became a hot scientific area that began to attract many researchers to the field. We also describe the first huge medical and economic impact that PDT as the first effective treatment for age-related macular degeneration, the leading cause of adult blindness in the world. Finally, we go through the main discoveries in veterinary medicine over the past years for the treatment of localized tumors and infections in diverse animal species.

The history of photomedicine can be traced back to ancient times when there were attempts to treat skin diseases with compounds that demonstrated specific therapeutic effects when exposed to sunlight. Historical documents from over 3000 years ago mention ancient Egyptians using substances extracted from plants combined with sunlight to treat superficial lesions. In Egypt, India, and China, sunlight was used to treat skin diseases such as psoriasis, vitiligo, and cancer. These practices are described in ancient texts such as the *Ebers Papyrus* [1] and the sacred Indian book, *Atharva Veda* [2].

Since the early 1800s, the scientific community was aware that ultraviolet and blue light were the best colors (wavelengths remained unknown during this period) to induce chemical reactions; for example, the darkening of silver nitrate led to the development of photography, even though it did not induce heating effects. On the other side of the spectrum, red and infrared light behaved in the opposite way, producing heat but few chemical reactions. As such, ultraviolet and blue light became commonly referred to as "chemical rays," while red and infrared light were called "thermal rays." In 1901 Niels Finsen, a Danish medical scientist, published the first modern book about the treatment of diseases with monochromatic light entitled, *La Phototherapie* (in English, *The Phototherapy*). He initially observed how chemical rays induced the most pronounced physiological effects in living cells during and after illumination. Finsen first made the key observation that blue light did not induce deleterious inflammatory responses, as did UV radiation, but could still kill bacteria. Based on this principle, he built many variations of optical systems to focus intense but non-thermally acting, blue light and noticed that some species of bacteria could be inactivated within minutes when illuminated. These methods and tools Finsen developed proved to be very effective to treat skin infections (Fig. 1.1). As a result, many clinics in Northern Europe adopted his techniques into their own practices. In 1903, Finsen was awarded the Nobel Prize in Physiology or Medicine "in recognition of his contribution to the treatment of diseases, especially cutaneous tuberculosis, with concentrated light radiation, whereby he has opened a new avenue for medical science" (*Nobel Prize Website*) [3, 4].

During this same period, the use of photosensitizers for therapeutic purposes also began to be investigated. The first report was made in 1900 by Prime, a French neurologist who was administering eosin dye orally for epilepsy treatment in humans. Apart from the neurologic effects, his trials suggested that skin regions exposed to sunlight developed severe erythema due to increased photosensitivity

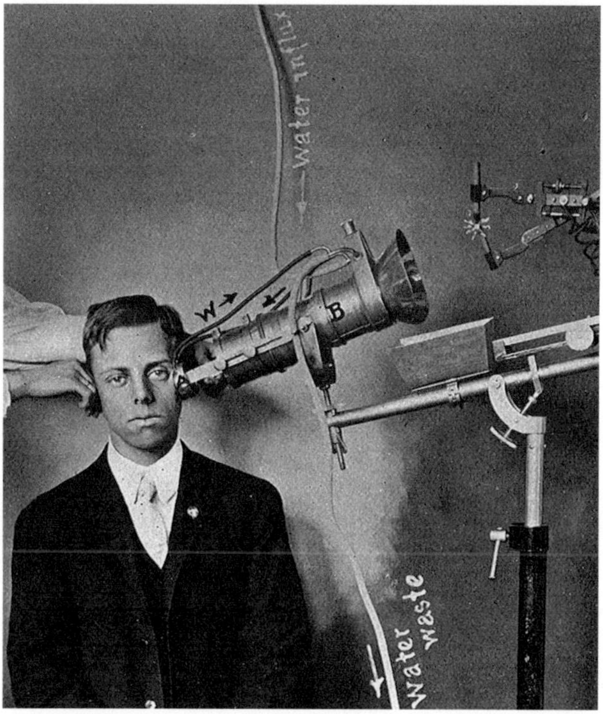

Fig. 1.1 Monochromatic light irradiation system developed by Niels Finsen for the treatment of cutaneous tuberculosis. The system consists of a carbon arc light source coupled to hollow glass lenses filled with blue dyes to filter focused blue light on patient skin lesions. A cooling system with circulating water was necessary to avoid overheating (Adapted from *La Phototherapie*)

[5]. Also in 1900, Oscar Raab, a German medical student under the supervision of Professor Herman Von Tappeiner, made the first scientific observation of photodynamic activity completely by accident. During his experiments on the viability of motile *protozoa*, Raab observed how fluorescent dyes like acridine orange, could kill stained microbes when sunlight was focused into his microscope. Such effects were more evident during summer, when sunlight is strongest. This simple observation demonstrated how some exogenous fluorescent compounds, now termed "photosensitizers" (PS), could be used to artificially induce light sensitivity in microorganisms and improve the already-known antimicrobial activity of sunlight [6]. Soon after his first discovery, Von Tappeiner, together with Jodlbauer, found that oxygen was necessary to develop reactions mediated by light, thereby creating the term *photodynamische wirkung* (photodynamic effect) [3, 4, 7–9]. At the same time, Von Tappeiner and Jesionek investigated the application of eosin and light for treatment of skin tumors in mice [10–12]. Despite the promising results they were obtaining, PDT was abandoned because the photosensitizers they used (dyes such as acridine) were found to be too toxic and presented a high oncogenic potential [5, 13].

In 1910, Hausmann made a tremendous breakthrough: he demonstrated the ability to destroy the tumor vasculature in mice following administration of hematoporphyrin (isolated from cattle blood) and exposure to sunlight [14]. In contrast to

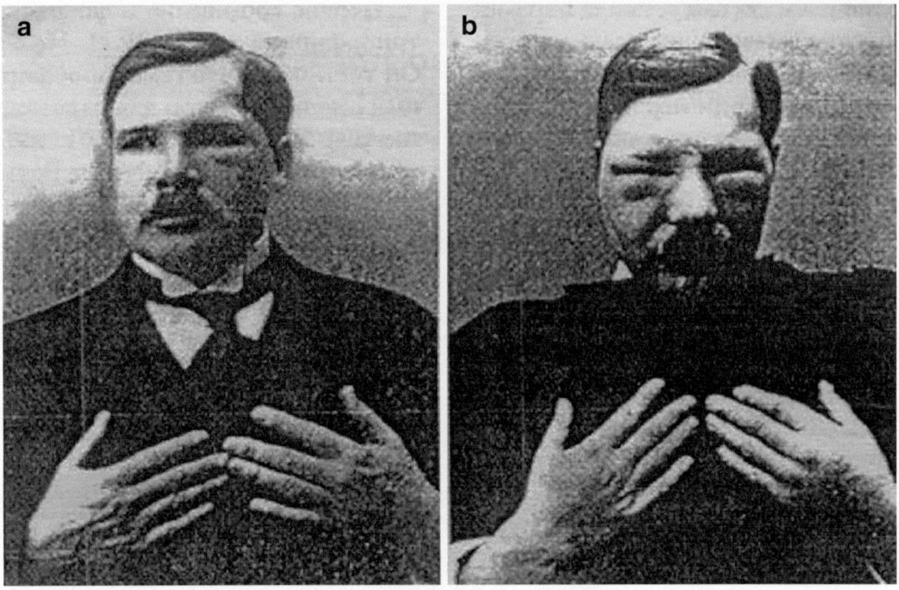

Fig. 1.2 Photograph of Friedrich Meyer-Betz, in 1913. On the *left* (**a**), we observe the scientist before hematoporphyrin administration. On the *right* (**b**), it is possible to observe hematoporphyrin-related effects after sun exposure [15]

eosin and acridines, porphyrins were not toxic in the absence of light. In 1913, Friedrich Meyer-Betz injected himself with 200 mg of purified hematoporphyrin and stayed under sunlight for a few minutes. No drug-related effects were noticed prior to sunlight exposure; however, upon sun exposure, his skin became extremely swollen and reactions were observable for months (Fig. 1.2) [15].

Years later, in 1924, Policard detected fluorescence emitted from tumors after intravenous administration of porphyrins followed by illumination with black light (UVA about 360 nm). Initially, it was believed that fluorescence was related to porphyrins occurring in secondary infections commonly found in some tumor lesions. There was no further research in this field until 1942 by scientists Auler and Banzer. The two made systematic studies reporting how malignant tumors indeed emitted fluorescence after systemic administration of porphyrins because these molecules preferentially accumulate in tumors [6, 16, 17].

Concurrently with studies investigating the action of PDT, in 1916 Albert Einstein postulated "the principles of light amplification by stimulated emission of radiation" that provided the theoretical basis for the future development of lasers. The first working laser was built in 1960 by Theodore Maiman. This development triggered a series of research studies involving the interaction of intense and monochromatic light with living organisms [18]. Until then, light sources available for PDT were the sun and enormous electric light systems such as carbon arc or pointolite lamps, which could provide noncoherent polychromatic light that required complex sets of filters and efficient water-cooled heat sinks for practical use [6].

The first generation of photosensitizers based on hematoporphyrin derivatives began in 1950 with studies led by Schwartz. The researchers demonstrated how in Meyer-Betz's experiments the most active component was not hematoporphyrin itself but was actually a mixture of oligomeric porphyrins isomers not easily eliminated from the body. Understanding the need for improvements, Schwartz optimized this mixture of oligomers calling it hematoporphyrin derivative (HPD), which was much more efficient than the substance used by Meyer-Betz. Later in the 1960s, Schwartz and Lipson demonstrated the preferential accumulation of HPD in tumors. These observations led to the development of many studies designed to detect tumors by endoscopic fluorescence detection [19]. In the latter half of the 1960s, Lipson successfully treated a woman with breast cancer using HPD and selective tumor irradiation, marking the beginning of oncologic PDT [20].

In the 1970s, Thomas Dougherty initiated a series of clinical trials of anticancer PDT using HPD [12]. He led efforts to develop feasible formulations and a large-scale production method to meet the standards set by the Food and Drug Administration (FDA) [11]. The 1970s and 1980s were marked by numerous studies regarding PDT for cancer therapy; however, antimicrobial PDT hardly attracted any attention, since antibiotics had been introduced in the 1940s and was still highly effective against infectious diseases. The discovery of penicillin by Alexander Fleming in 1928 opened the path for antimicrobial chemotherapy and the following decades became known as the *age of antibiotics*. This pharmacologic approach was markedly more effective and versatile than PDT to treat local and systemic infections. In addition, antibiotics were much more profitable and could easily be transported and stored in remote locations (this was especially important during times of war). But as a result of indiscriminate use of antibiotics in the last few decades, the selection of resistant populations of microorganisms has revived an urgent need for therapeutic alternatives. This has led to antimicrobial PDT beginning to regain attention for the treatment of infectious disease. Over the last decade, several photosensitizer candidates with diverse molecular frameworks have been tested to inactivate all sorts of pathogenic microorganisms (different classes of bacteria, fungi, viruses, and parasites). Due to their broad-spectrum activity, low-cost, and easy availability, phenothiazinium dyes such as methylene blue and toluidine blue became the most commonly used PS for microbial inactivation [21]. The twenty-first century is marked by increasing interest in antimicrobial PDT with extensive development of preclinical and clinical trials that endure to this day. As mentioned above, bacteria, fungi, protozoa, and viruses have been inactivated with various degrees of success depending on experimental conditions [22] (Fig. 1.3).

The history of antimicrobial PDT may have started when Oscar Raab accidentally inactivated paramecia in 1900. However, after this first observation, all research was focused on cancer therapy. Only in 1931 Clifton published the first study demon-

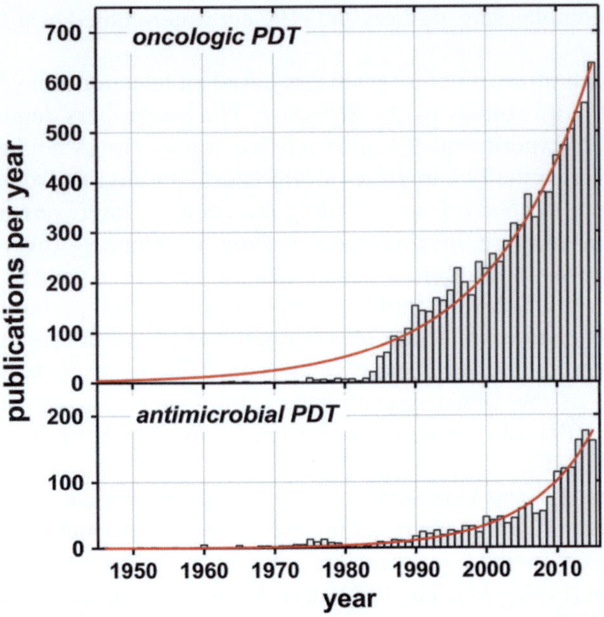

Fig. 1.3 Number of indexed publications per year over the past six decades according to PUBMED database

strating the antiviral activity of PDT using phenothiazine dyes and sunlight to inactivate bacteriophages [23]. Based on this discovery, Perdrau and Todd developed a highly antigenic vaccine against canine distemper with viral particles inactivated by methylene blue and artificial light of a pointolite lamp [24]. The use of PDT to treat active human infections only began to be reported in 1973 by a research group lead by Joseph Melnick. In the same year, they published two double-blind clinical studies reporting beneficial effects on management of herpes simplex lesions in various anatomic sites [25, 26]. In addition to shorter healing time, recurrences were also reduced in relation to placebo controls. Unfortunately, a former member of Melnick's lab falsely reported to the press that photodynamically treated herpesvirus might cause cancer but omitted that herpes lesions alone lead to cancer predisposition [27]. Such affirmation obviously generated a great confusion among the medical community, and most physicians were discouraged to use PDT against infections. Now that PDT has been used in clinics for many years, we have solid scientific and historical basis to affirm that oncogenic side effects are very unlikely to occur.

Possibly the most financially successful application in all of PDT was neither its use for cancer nor its use for localized infections. Instead, this application involved the use of PDT in ophthalmology to prevent and treat the most common cause of adult blindness. This disorder originated in what is known as "age-related macular degeneration (AMD)," a disease involving the buildup of lipid deposits (drusen) in the cells of the retinal-pigmented epithelium. These deposits cause chronic inflammation that eventually leads to atrophy and scarring of the retina. A more serious consequence is the "wet form of AMD" where abnormal blood vessels grow (choroidal neovascularization) in the choriocapillaris. The proliferation of abnormal blood vessels in the retina is stimulated by vascular endothelial growth factor (VEGF). These new vessels are leaky and bleeding, and scarring from these blood vessels eventually causes irreversible damage to the photoreceptors and rapid vision loss if left untreated [28]. A systemic injection of benzoporphyrin derivative (BPD) also known as Visudyne or verteporfin, followed by 10 min of irradiation to the back of the eye with a 690 nm laser, destroyed these abnormal vessels [29] and was shown to prevent further deterioration and even improve eyesight [30]. For several years, this approach was the treatment of choice for wet AMD, until it was to some degree supplanted by intravitreous injection of anti-VEGF antibodies [31].

The introduction of PDT into veterinary medicine took place in the 1980s, when several types of tumors in dogs and cats were treated with HPD and irradiated with laser light. Although these studies only addressed a limited number and variety of animals and tumor types, most lesions responded well to treatment under established protocols [32–36]. So far, the search for effective alternatives for treatment of cancer and infectious diseases has been performed on pets, farm animals, and wild animals [37, 38]. Despite the great potential of PDT to treat diseases caused by different etiological agents, only a limited number of studies have been reported so far in veterinary medicine when compared to other areas such as oncology and dentistry. However, based on the growing market of specialized therapeutic interventions for pets and livestock, along with the ecological appeal of wildlife conservation, efficient alternatives to conventional treatments such as surgery and chemotherapy have become of paramount importance. Figure 1.4 summarizes a historic timeline of PDT highlighting the most relevant milestones for PDT in veterinary sciences.

Acknowledgments MR Hamblin was supported by US NIH grant R01AI050875.

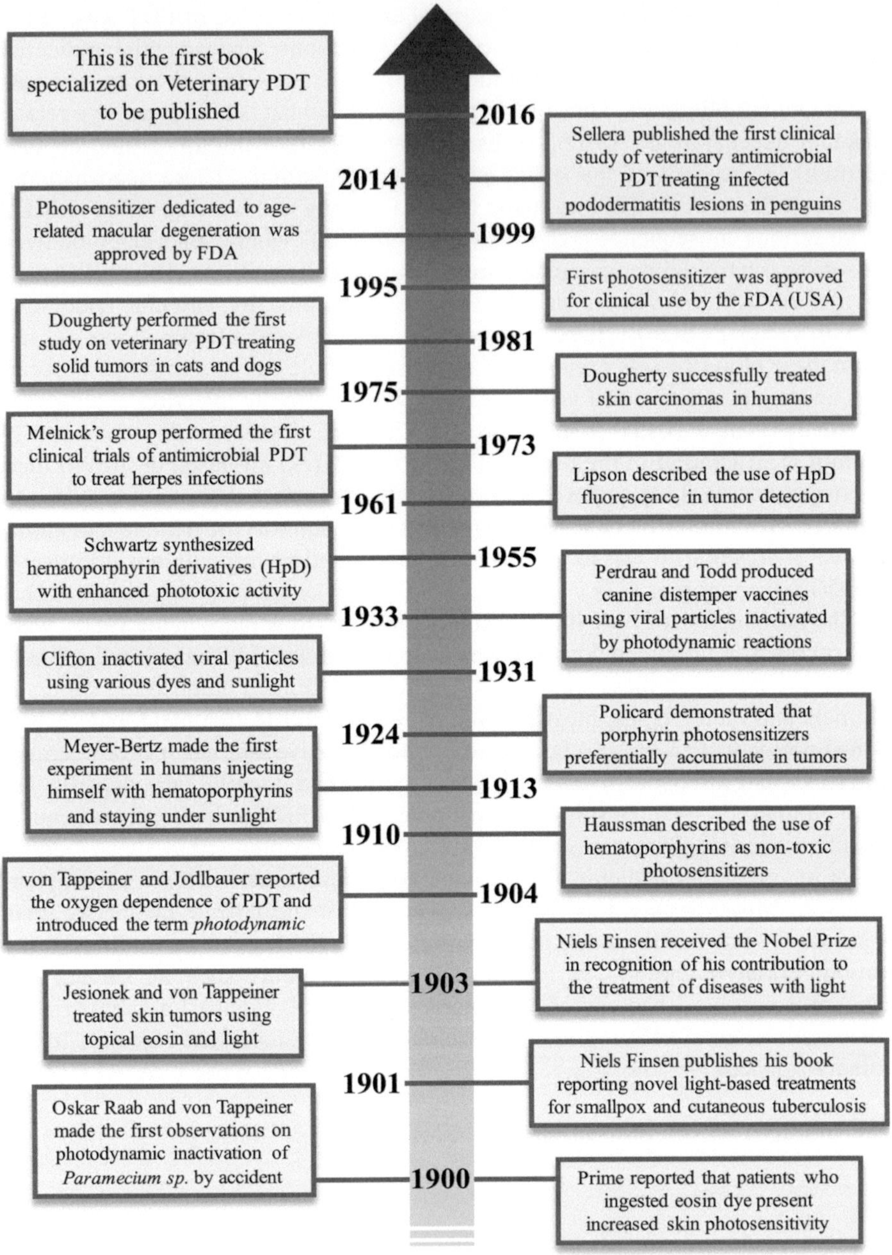

Fig. 1.4 Timeline of PDT highlighting the most relevant milestones for PDT in veterinary sciences

References

1. El-Mofty AM. Clinical experience in the treatment of vitiligo with psoralens. In: Vitiligo and psoralens. 1st ed. Oxford: Pergamon Press; 1968. p. 147.
2. Wyss P. History of photomedicine. In: Wyss P, Tadir Y, Tromberg BJ, Haller U, editors. Photomedicine in gynecology and reproduction. 2nd ed. Karger Publishers, Basel, Switzerland; 2000. p. 4–11.
3. Peng Q, Moan J, Nesland JM. Correlation of subcellular and intratumoral photosensitizer localization with ultrastructural features after photodynamic therapy. Ultrastruct Pathol. 1996;20(2):109–29.
4. Allison RR, Downie GH, Cuenca R, Hu X-H, Childs CJ, Sibata CH. Photosensitizers in clinical PDT. Photodiagnosis Photodyn Ther. 2004;1(1):27–42.
5. Prime J. Les accidentes toxiques par l'eosinate de sodium. Paris: Jouve and Boyer; 1900.
6. Ackroyd R, Kelty C, Brown N, Reed M. The history of photodetection and photodynamic therapy. Photochem Photobiol. 2001;74(5):656–69.
7. Von Tappeiner H, Jodlbauer A. Uber Wirkung der photodynamischen (fluorieszierenden) Stoffe auf Protozoan und Enzyme. Dtsch Arch Klin Med. 1904;80:427–87.
8. Von Tappeiner H, Jodlbauer A. Die Sensibilisieren-de Wirkung fluorieszierender Substanzer. Gesammte Untersu-chungen uber die photodynamische Erscheinung. Leipzig: F. C. W. Vogel; 1907.
9. Babilas P, Karrer S, Sidoroff A, Landthaler M, Szeimies R-M. Photodynamic therapy in dermatology – an update. Photodermatol Photoimmunol Photomed. 2005;21(3):142–9.
10. Von Tappeiner H, Jesionek A. Therapeutische Ver-suche mit fluoreszierenden Stoffen. Muench Med Wochenschr. 1903;47:2042–4.
11. Kessel D. Photodynamic therapy: from the beginning. Photodiagnosis Photodyn Ther. 2004;1(1):3–7.
12. Dougherty TJ, Kaufman JE, Goldfarb A, Weishaupt KR, Boyle D, Mittleman A. Photoradiation therapy for the treatment of malignant tumors. Cancer Res. 1978;38(8):2628–35.
13. Sternberg ED, Dolphin D. Porphyrin-based photosensitizers for use in photodynamic therapy. Tetrahedron. 1998;54:4151–202.
14. Hausmann WH. Die sensibilisierende Wirkung des Hamatoporphyrins//Biochem Z. Biochem Z. 1910;30:276–316.
15. Meyer-Betz F. Untersuchungen über die biologische (photodynamische) Wirkung des Hämatoporphyrins und anderer Derivate des Blut- und Gallenfarbstoffes. Dtsch Arch Klin Med. 1913;112:476–503.
16. Auler H, Banzer G. Untersuchungen über die Rolle der Porphyrine bei geschwulstkranken Menschen und Tieren. Z Krebsforsch. 1942;53(65):68.
17. Figge FHJ, Weiland GS, Manganiello LOJ. Cancer detection and therapy; affinity of neoplastic, embryonic, and traumatized tissues for porphyrins and metalloporphyrins. Proc Soc Exp Biol Med. 1948;68(3):640.
18. Spyropoulos B. 50 years LASERS: in vitro diagnostics, clinical applications and perspectives. Clin Lab. 2011;57(3–4):131–42.
19. Lipson RL, Baldes EJ, Olsen AM. The use of a derivative of hematoporphyrin in tumor detection. J Natl Cancer Inst. 1961;26:1–11.
20. Lipson RL, Gray MJ, Baldes EJ. Hematoporphyrin derivative for detection and management of cancer. In: Proceedings of the 9th International Cancer Congress. Tokyo; 1966. p. 393.
21. Wainwright M. Photodynamic antimicrobial chemotherapy (PACT). J Antimicrob Chemother. 1998;42(1):13–28.
22. Hamblin MR, Hasan T. Photodynamic therapy: a new antimicrobial approach to infectious disease? Photochem Photobiol Sci. 2004;3(5):436–50.
23. Clifton CE. Photodynamic action of certain dyes on the inactivation of staphylococcus bacteriophage. Exp Biol Med. 1931;28(7):745–6.

24. Perdrau JR, Todd C. Canine distemper. The high antigenic value of the virus after photodynamic inactivation by methylene blue. J Comp Pathol Ther. 1933;46:78–89.
25. Felber TD. Photodynamic inactivation of herpes simplex. JAMA. 1973;223(3):289.
26. Kaufman RH, Gardner HL, Brown D, Wallis C, Rawls WE, Melnick JL. Herpes genitalis treated by photodynamic inactivation of virus. Am J Obstet Gynecol. 1973;117(8):1144–6.
27. Melnick JL, Wallis C. Photodynamic inactivation of herpes simplex virus: a status report. Ann N Y Acad Sci. 1977;284:171–81.
28. Nowak JZ. Age-related macular degeneration (AMD): pathogenesis and therapy. Pharmacol Rep. 2006;58(3):353–63.
29. Messmer KJ, Abel SR. Verteporfin for age-related macular degeneration. Ann Pharmacother. 2001;35:1593–8.
30. Yang YC. Preserving vision with verteporfin photodynamic therapy. Hosp Med. 2004;65(1):39–43.
31. Yu J, Ba J, Peng R, Xu D, Li Y, Shi H, et al. Intravitreal anti-VEGF injections for treating wet age-related macular degeneration: a systematic review and meta-analysis. Drug Des Dev Ther. 2015;9:5397.
32. Dougherty TJ, Thoma RE, Boyle DG, Weishaupt KR. Interstitial photoradiation therapy for primary solid tumors in pet cats and dogs. Cancer Res. 1981;41(2):401–4.
33. Thoma RE, Stein RM, Weishaupt KR, Dougherty T. Phototherapy: A promising new cancer therapy. Vet Med Small Anim Clin. 1983;78:1693.
34. Cheli R, Addis F, Mortellaro CM, Fonda D, Andreoni A, Cubeddu R. Hematoporphyrin derivative photochemotherapy of spontaneous animal tumors: clinical results with optimized drug dose. Cancer Lett. 1984;23(1):61–6.
35. Cheli R, Addis F, Mortellaro CM, Fonda D, Cubeddu R. Photodynamic therapy of spontaneous animal tumors using the active component of hematoporphyrin derivative (HPPH) as photosensitizing drug: clinical results. Cancer Lett. 1987;38(1–2):101–5.
36. Lucroy MD. Photodynamic therapy for companion animals with cancer. Vet Clin North Am Small Anim Pract. 2002;32(3):693–702.
37. Sellera FP, Sabino CP, Ribeiro MS, Fernandes LT, Pogliani FC, Teixeira CR, et al. Photodynamic therapy for pododermatitis in penguins. Zool Biol. 2014;33(4):353–6.
38. Buchholz J, Walt H. Veterinary photodynamic therapy: a review. Photodiagnosis Photodyn Ther. 2013;10(4):342–7.

Chapter 2
Photophysical and Photochemical Mechanisms

Caetano Padial Sabino and Michael Richard Hamblin

Abstract Photodynamic therapy (PDT) harnesses the power of light in an elegant method to produce cytotoxic agents in a spatially and temporally controlled manner and specifically damage target cells and tissues. For photodynamic reactions to occur, the PS molecule must absorb at least one photon to be promoted to a sufficiently long-lived excited state and then induce photodynamic reactions in an oxygenated environment. Such properties guarantee that PDT has an exceptionally broad action spectrum against tumors or pathogens, and resistance occurrence is restricted to only a few exceptions that can be avoided using simple strategies. To fully understand the intricacies of the mechanisms by which PDT acts, it is clear that one must take advantage of all the basic sciences (e.g., physics, chemistry, and biology). In fact, such conceptual complexity still maintains constant scientific investigations to deeply understand the molecular basis of PDT. Curiously, it might also be one of the reasons to explain why this hundred-year-old technique is still not generally applied in clinics or taught in standard courses of pharmacology. In this

C.P. Sabino (✉)
Department of Microbiology, Institute for Biomedical Sciences, University of São Paulo, Av. Lineu Prestes 1347, Cidade Universitária, Sao Paulo 05508-000, SP, Brazil

Department of Clinical Analysis, School of Pharmaceutical Sciences, University of São Paulo, Sao Paulo, SP, Brazil

Center for Lasers and Applications, Nuclear and Energy Research Institute, National Commission for Nuclear Energy, Sao Paulo, SP, Brazil
e-mail: caetanosabino@gmail.com

M.R. Hamblin
Wellman Center for Photomedicine, Massachusetts General Hospital, 50 Blossom Street, Bartlett Hall, Room 414, Boston, MA, USA

Department of Dermatology, Harvard Medical School, Boston, MA, USA

Harvard-MIT Division of Health Sciences and Technology, Cambridge, MA, USA
e-mail: hamblin@helix.mgh.harvard.edu

F.P. Sellera et al. (eds.), *Photodynamic Therapy in Veterinary Medicine: From Basics to Clinical Practice*, DOI 10.1007/978-3-319-45007-0_2

chapter, we will attempt to use a multidisciplinary approach, with simple technical language and a minimum of mathematics and equations, to allow any student with minimal training in basic sciences to understand all the fundamental mechanisms of PDT.

2.1 Introduction

Ever since the very first moments following the Big Bang cosmological event, electromagnetic radiation (EMR) has spread through the whole universe introducing various degrees of perturbations into the properties of condensed matter. Our planet Earth was formed and evolved continuously receiving large amounts of radiation from our closest star, the Sun. Human kind has recognized long ago that life is impossible without sunlight and in diverse historical periods has worshiped the Sun, or a personification of it, as a powerful god. During daylight primitive people used light to search for food and escape from predators, and during the long and warm days of summer, living became easier and food was much more abundant. In fact, current scientific knowledge proposes that all life on the Earth evolved under the influence sunlight, which induced the formation of the very first organic molecules in the atmosphere or oceans [1].

At some point, living organisms developed the capacity to use photosynthesis, which became a complex series of physicochemical reaction processes to harness energy from sunlight to produce the basic organic compounds needed for cellular metabolism. Over billions of years, all life thrived in such a "radiation-rich" environment forming a plethora of variations in the food chain all of which depended on photosynthesis either directly or indirectly. However, photosynthesis is far from being the only physicochemical processes based on absorption of energy from light that life-forms have evolved to relay on. Light also represents the means we use to interact with the environment (e.g., different forms of vision), regulates our circadian rhythms through night and day cycles, and catalyzes the production of micronutrients (e.g., vitamin D). Light can be a source of cellular damage (e.g., erythema induced by ultraviolet and DNA damage caused by X-rays, gamma rays, and UV and even thermal effects caused by infrared radiation), but light can also cause the repair of damage to biomolecules and tissues (e.g., photolyase enzyme, photobiomodulation therapy).

Pigmentation seems to play an important role in the animal kingdom as a protective defense against radiation. It is well known that solar erythema affects animals, such as cattle and horses, almost exclusively in the lightly pigmented areas, while the darker parts are less affected. Hence, the anatomical regions most exposed to the sun are usually more deeply pigmented than less sun-exposed regions. Examples of this are found almost everywhere: among furred animals, whales, reptiles, birds, fish, etc. These phenomena are not very different from what we see in humans. Due to geographic dispersal during human evolution, populations inhabiting regions close to the equator present more strongly pigmented skin than those who evolved in regions closer to the polar regions. Yet, people with darker skin tones still show much lighter pigmentation in regions less exposed to the sun, such as the palms of hands and soles of the feet.

Nowadays it is well known that some wavelengths, or "colors", of visible and near-ultraviolet (NUV) light can damage and even kill bacteria. In the late 1800s, researchers performed several investigations and concluded that sunlight was the best, cheapest, and most universally applicable antimicrobial agent we have [2]. Back then, the NUV and blue regions (300–500 nm) of the EMR spectrum were referred as chemical rays because the chemical effects were more pronounced and the heating effects were at a minimum. Using dyes, crystals, and prisms to separate light from the sun into monochromatic colors, the so-called chemical rays were determined to be the most effective in killing bacteria and causing erythema than the opposite end of the spectrum, the infrared "thermal rays." It was then realized that cellular damage and inactivation induced by sunlight was more closely related to photochemical reactions than photothermal effects. However, the photochemical mechanisms behind such effects would only be understood about half a century later, even though Niels Finsen was awarded a Nobel Prize in the meantime (see Chap. 1). Currently, it is known that EMR from sunlight can directly damage biomolecules through ionization (e.g., production of free radicals, photodimerization of thymine residues in DNA) but also acts indirectly through excitation of endogenous photosensitizers (PS) (e.g., riboflavin and metal-free porphyrins) present in cells producing reactive oxygen species (ROS) through photodynamic reactions [3].

Photodynamic therapy (PDT) harnesses the power of light in an elegant method to produce cytotoxic agents in a spatially and temporally controlled manner and specifically damage target cells and tissues. For photodynamic reactions to occur, the PS molecule must absorb at least one photon to excite its electronic state as predicted by quantum theory. PS compounds with a sufficiently long-lived excited state can then induce photodynamic reactions that are directly or indirectly dependent on the presence of O_2 to locally form free radicals and singlet oxygen ($^1O_2$). Since many PS are nontoxic in the dark and can preferentially accumulate in tumors or pathogens, only minor side effects are expected. However, when the PS-loaded cells are illuminated, multiple relevant biomolecules can be rapidly degraded. Such properties guarantee that PDT has an exceptionally broad action spectrum against tumors or pathogens, and resistance occurrence is restricted to only a few exceptions that can be avoided using simple strategies.

To fully understand the intricacies of the mechanisms by which PDT acts, it is clear that one must take advantage of all the basic sciences (e.g., physics, chemistry, and biology). In fact, such conceptual complexity is still a subject of study and might be one of the reasons to explain why this hundred-year-old technique is still not generally applied in clinics or taught in standard courses of pharmacology. In this chapter, we will attempt to use a multidisciplinary approach, with simple technical language and a minimum of mathematics and equations, to allow any student with minimal training in basic sciences to understand all the fundamental mechanisms of PDT.

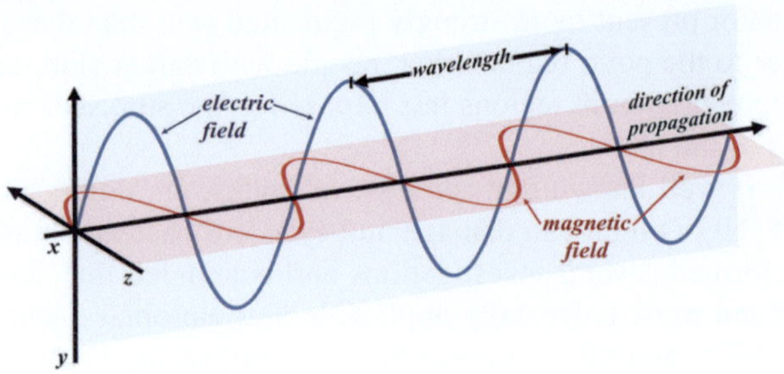

Fig. 2.1 Schematic illustration of a classic electromagnetic wave. It consists of transverse oscillations of electric and magnetic fields that are perpendicular to each other and in relation to the propagation (motion) direction. Wavelength is determined by the distance between two peaks or valleys. The speed of light in vacuum is $c_0 = 299,792,458$ m/s

2.2 Electromagnetic Radiation

Before presenting the biological effects of PDT, this chapter will introduce the nature of light and how it interacts with the electrons present in atoms and molecules to induce physical and chemical effects that will eventually lead to photodynamic reactions. The term electromagnetic radiation refers to a certain kind of energy that propagates through space (what used to be called the "ether"), and which can be interpreted similarly to heat being considered the propagation of vibrational energy through solid material, or wind as being the propagation of kinetic energy through air. However, this form of radiant energy, or radiation, can propagate in absolute vacuum at the highest speed theoretically possible in nature: the speed of light, $\mathbf{c_0} = 299,792,458$ meters per second (m/s). The question of whether EMR is better interpreted as a wave or as a particle (i.e., photon) has spurred exhaustive discussions among scientists, over the past few centuries. Only with the advent of quantum mechanics and precise experiments carried out over the last century has it been possible to demonstrate that both interpretations were equally valid for diverse situations. Depending on the physical observation, EMR can either be interpreted as a wave or as a particle carrying a discrete amount, or "quantum," of energy. In fact, Louis de Broglie postulated in 1924 the "wave-particle duality theory" proposing that any accelerated particle is associated with a wave, and vice versa (any wave is associated with a particle), thus redefining the whole concept of radiation itself [4]. Under his theoretical interpretation, the kinetic energy of any moving particle with mass, such as protons and electrons in particle accelerators, can be directly related to a certain wavelength of corpuscular radiation. For the purpose of this book, however, we will retain our attention only to massless and charge-free particles of EMR called "photons."

EMR is classically defined as transverse waves with oscillations of electric and magnetic fields that are perpendicular to each other and both perpendicular to the direction of propagation (Fig. 2.1). As any wave propagating at a certain velocity (V), EMR can be described in terms of specific wavelength (λ) and frequency (ν) of

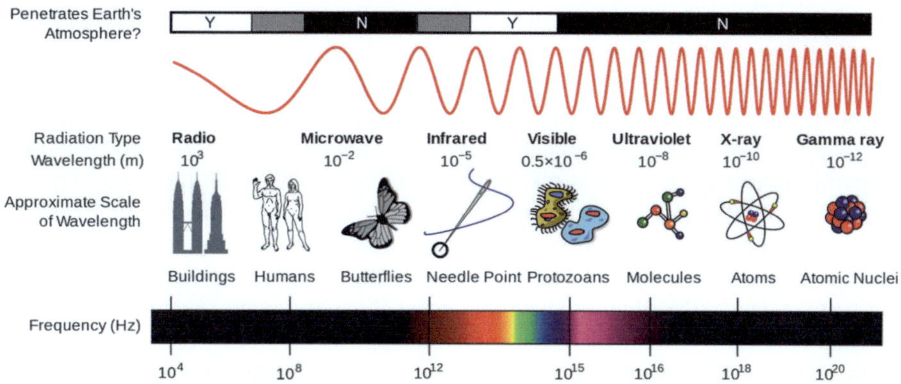

Fig. 2.2 Diagram of electromagnetic spectrum illustrating various properties across a broad range of wavelengths or frequencies. Note that only some regions of radio waves and visible light can transpose the current Earth's atmosphere and reach the ground (Image was modified from Inductive load, at Wikipedia. This image has "GNU Free Documentation License")

oscillation (Eq. 2.1). According to the international system of units, **λ** is measured in meters (m) and ν in Hertz (Hz, i.e., number of oscillations per second). In some circumstances, the wave number ($\bar{\omega}$, i.e., number of waves per centimeter) is alternatively used to characterize EMR. Modern physics describes EMR as a photon particle where the exact quantum of energy (***E***) of each individual photon can be easily calculated in respect to its frequency times the Planck constant ($\boldsymbol{h}=6.6260700\times10^{-34}$ J.s), as presented in Eq. (2.2). Hence, according to Eqs. (2.1 and 2.2), in lower wavelengths or higher frequencies, the photon energy becomes higher:

$$V = c_0 = \lambda \upsilon \tag{2.1}$$

$$E = h\upsilon = \frac{h.c_0}{\lambda} \tag{2.2}$$

The energy associated with EMR has a variation of many orders of magnitude across the whole spectrum (Fig. 2.2). According to increasing levels of frequency or energy, the electromagnetic radiation is classified into radio waves, microwaves, infrared radiation, visible light, ultraviolet radiation, X-rays, and gamma rays. According to Fig. 2.2, for example, one single gamma ray photon can carry up to 100,000 times the energy of a single visible light photon. Therefore, a gamma ray photon can induce physical phenomena in atoms or molecules that require 100,000 times the energy of a visible light photon. Conversely, a gamma ray photon would have 10^{15} times as much energy as a radio wave photon.

The physical phenomena that occur in matter associated with the emission or absorption of a photon are directly proportional to its energy. The less energetic radio- and microwave photons induce rotation of molecules and atoms to increase their kinetic energy as they spin around their center of mass. Electricity-conducting materials, such as metals, also allow electron translation and produce electric currents. Infrared radiation, at wavelengths from 1 to 1000 μm, increases the kinetic energy of molecules by means of vibration. The chemical bounds within molecules are not rigid. They behave mostly as flexible springs constantly oscillating at vibrational modes of certain frequencies and coordination (e.g., stretching, bending, wagging, twisting, etc.). Each different type of molecular bond has characteristic vibrational modes that can be promoted by infrared photons of matching energy. Since temperature is a measure of the average kinetic energy of particles of matter, such as atoms and molecules, the absorption of these types of radiation by biological systems is mostly related to heat induction alone and often leads to nonspecific thermal effects on biological systems. On the opposite extremity of the EMR spectrum, there are X-rays and gamma rays with very high-energy photons that remove electrons from inner electronic shells of atoms or molecules and produce free radicals and ions. We commonly refer them as ionizing radiation due to this strong characteristic. This radiation can severely damage living organisms by direct and indiscriminate ionization of biomolecules, allied to simultaneous formation of free radicals originating from broken bonds. Ionized molecules become extremely reactive, and chemical bonds may be consequently broken or formed. Since absorption of these high-energy photons is generally nonspecific, the ability to "target" molecules is poor and constantly leads to the damaging side effects of radiotherapy.

Only a small portion of the EMR spectrum can be used to target the excitation (but not removal) of electrons in specific molecules with satisfactory precision. This region encompasses near-UV, visible, and near-infrared radiation (i.e., λ approximately from 300 to 900 nm). Since photodynamic reactions are governed by absorption of photons within this range, in this book the term "light" will always be used to mean this range of the EMR spectrum. Differently from other types of radiation we mentioned before, light interacts with electrons from valence shells of molecules or atoms that often absorb very characteristic wavelengths. As an example of such specificity, all vivid colors we can see are within the tiny interval of whole EMR spectrum known as visible radiation. A combination of red (600–700 nm), green/yellow (500–600 nm), and blue (400–500 nm) light is detected by retinal cones in our eyes and interpreted by our brain. Color-blind individuals are unable to see one or more of these colors because some of their retinal cones are absent or defective. Since what we detect is visible light that was either reflected, refracted, or emitted by some material, all color differences are relative to light absorption and scattering properties of each material. Take plant chlorophyll as an example: when white sunlight interacts with leafs, red and blue light are strongly absorbed, while all the green light is reflected and can be detected by our eyes. Note in Fig. 2.3 how small molecular alterations, such as a carboxyl group addition (red arrow), can shift light absorption bands among chlorophylls *a* and *b*. Chlorin e6 is a chlorophyll derivative molecular framework used as PS for PDT. The exclusion of the lipid radical and chelated Mg^{++} in chlorin e6 introduces further absorption shifts in relation to chlorophylls.

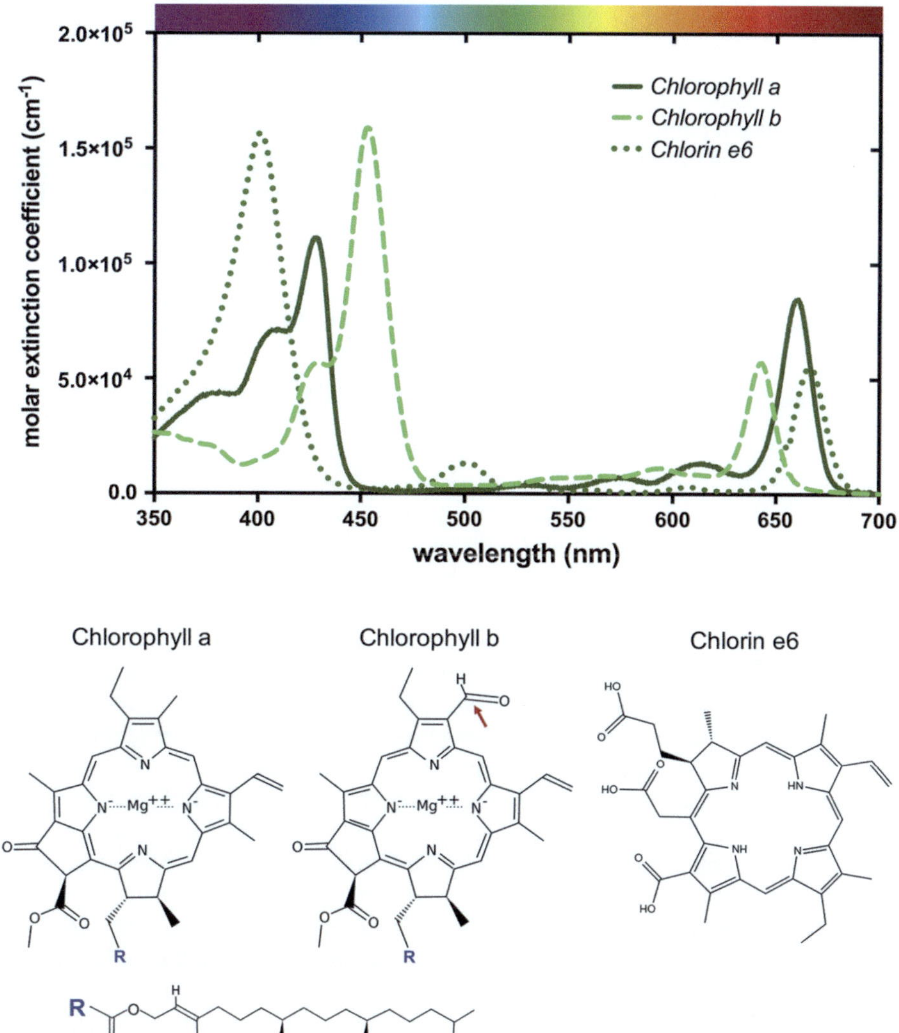

Fig. 2.3 Absorption spectra of three *green* pigments derived from plants: chlorophyll a, chlorophyll b, and chlorin (e6) (a chlorophyll derivative used as a photosensitizer for PDT). Absorption intensity is presented as molar extinction coefficient units in vertical axis, and wavelength is plotted in horizontal axis. *Colored bar* at the *top* illustrates the regions of visible light spectrum. Below are presented their respective molecular structures. Observe how small molecular alterations, such as a carbonyl group addition (*red arrow*), can shift light absorption bands among chlorophyll a and b (Data acquired from omlc.org/spectra/PhotochemCAD/index.html [13])

2.3 Photophysics and Photochemistry

The first law of photochemistry states that light must be absorbed by a molecule in order to induce a photochemical reaction. The second law of photochemistry completes the concept of light absorption, stating that each photon absorbed by a chemical system can only excite a single molecule. Therefore, light absorption is a fundamental process that must be understood by those who intend to explore the world of PDT. Each photosensitive molecular structure possesses a characteristic probability of photon absorption at different regions of the EMR spectrum. This is commonly called an "absorption spectrum." The Lambert-Beer law proposes a simple exponential equation (2.3) to measure the monochromatic light absorption efficiency of photosensitive molecules dissolved in a transparent solvent:

$$I = I_0 \times 10^{-\varepsilon l c} \tag{2.3}$$

where $\boldsymbol{I_0}$ is the initial light beam intensity and $\boldsymbol{I}$ is the light intensity transmitted through a medium of path length $\boldsymbol{l}$ with an absorbing compound dissolved at concentration $\boldsymbol{c}$. The term $\boldsymbol{\varepsilon}$ is a characteristic constant of each compound, named the "molar extinction coefficient," that tells us how efficient a molecule is to absorb light at that wavelength (usually the wavelength of peak absorption λ_{max}). Efficient PS for PDT commonly have absorption peaks with $\boldsymbol{\varepsilon}$ values greater than 10,000 $M^{-1}cm^{-1}$. This quantity means that light intensity at a certain wavelength can be decreased by 10,000 times after propagating through 1 centimeter of this photosensitizer solution at 1 molar of concentration. Even though light absorption during PDT applications can be enhanced by increasing the PS concentration, this approach must be taken carefully since high concentrations of PS can be toxic in the dark and lead to undesirable side effects or diminished therapy specificity. Moreover, very high local concentrations of PS can reduce light transmission into the tissue.

When a "ground-state molecule" (i.e., not an excited molecule) absorbs one photon of light, one valence shell electron can be promoted from the highest occupied molecular orbital (HOMO) to the lowest unoccupied molecular orbital (LUMO) to form an "excited-state molecule." To absorb a photon, the energy gap between the ground- and excited-state molecule (HOMO-LUMO gap) must match the quantum of energy carried by the photon. The extra amount of energy brought by a photon significantly alters the electronic configuration of valence shells transforming the excited-state molecules into new chemical species that have their own physical and chemical properties, which can be very different from those of ground-state molecules. Excited-state molecules, however, are usually unstable with a very short lifetime and may quickly return to their ground state via various possible deactivating processes (Fig. 2.4). Those can be photophysical processes (with or without photon emission), or photochemical reactions leading to production of new molecular species, such as ROS generated by photodynamic reactions. Each photosensitive molecule has characteristic probability to undergo each process, and it can be influenced by solvents and other molecules present in its surroundings.

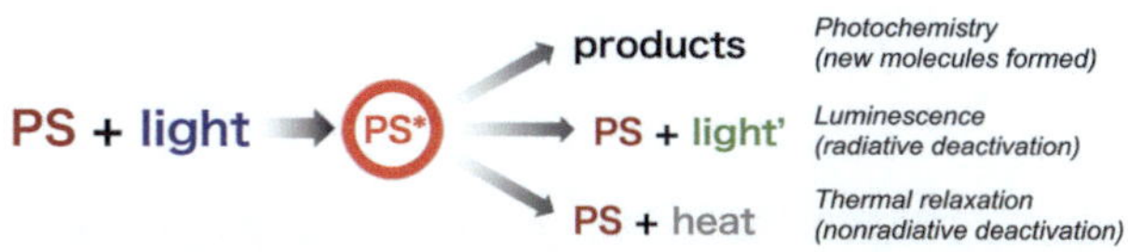

Fig. 2.4 Self-deactivation processes of photosensitizer (PS) molecules excited by light. Photochemical processes form new molecular species through isomerization, cleavage, synthesis, and electron or energy donation. The last two are predominant in photodynamic reactions. Photophysical processes can lead to luminescence through fluorescence or phosphorescence or heat generation through thermal relaxation (i.e., energy is transformed into kinetic energy)

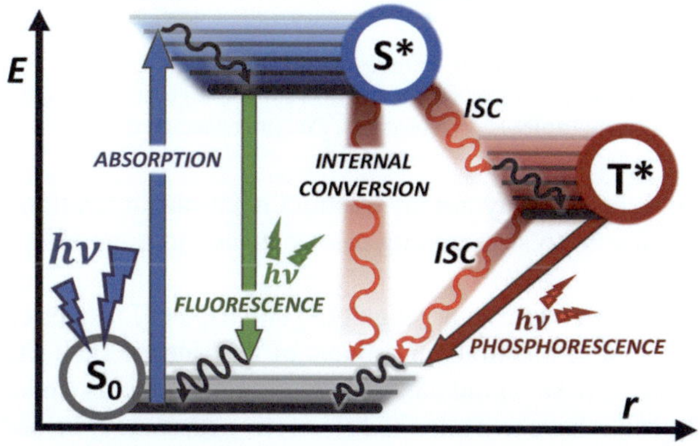

Fig. 2.5 Jablonski diagram illustration of possible photophysical processes. *Straight lines* represent radiative transitions and wavy lines nonradiative transitions. After photon absorption, ground-state singlet molecule is promoted to excited-state singlet and may decay back to ground state through fluorescence or internal conversion. Alternatively, S* may decay to a more stable triplet state through intersystem crossing (ISC) and only then be deactivated through phosphorescence or another ISC

There is a thin line that separates the concepts of photophysical and photochemical phenomena. To put it simply, photophysicists mostly deal with molecular and atomic excitation processes, including photon absorption itself, that lead to photon emission or thermal effects, but in general they do not focus on chemical reactions produced by excited states or the formation of new ground-state species. Of course, nowadays, many scientists work in multidisciplinary surroundings and researchers must consider all perspectives.

The "Jablonski diagram" (Fig. 2.5) is often used to conveniently illustrate such phenomena. It is can be interpreted as a map of physical and chemical events related to the interaction of light that can occur with a molecule. The electronic states are vertically separated in groups of parallel lines to illustrate the relative energies. Each line represents a particular vibrational mode of each state. In between the lines, there are multiple energy levels of rotational states. Radiative transitions, either by absorption or emission of a photon (***hν***), are represented in straight lines and nonradiative transitions as wavy lines. Excited-state molecules are indicated by asterisks (*). When a molecule is excited by light, an electron is promoted to higher

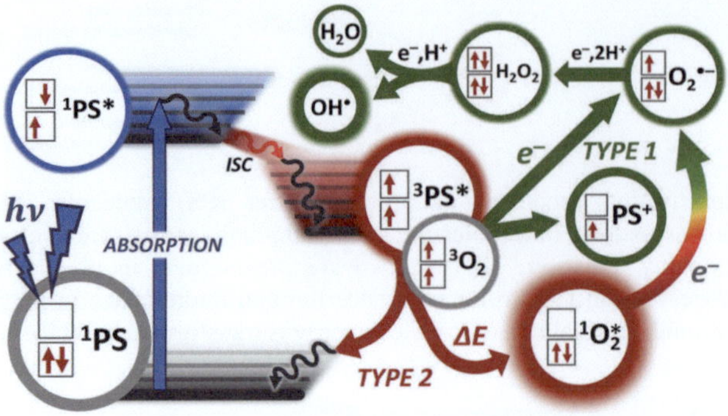

Fig. 2.6 Schematic illustration of main transitions and reactions directly involved with photodynamic reactions. *Arrows* in *squares* represent electron spin configurations in valence shell orbitals of radicals and singlet- and triplet-state molecules. *ISC* intersystem crossing

energy configurations and may assume distinct positions in relation to the nucleus due to altered repulsion to surrounding electrons, as represented by the horizontal axis in Fig. 2.5.

The definition of free radicals or singlet (S) and triplet (T) states can be based on the presence and spin of electrons in orbitals of the valence shell. As illustrated in Fig. 2.6, singlet states have paired electrons with opposite spins (indicated by arrow direction), triplet states have two unpaired electrons with same spin, and radicals have one unpaired electron. Therefore, when an excited singlet state (S*) decays to a triplet state, the electron spin is inverted (flipped). Since excitation to a triplet state involves an additional (forbidden) spin transition, it is less probable that a triplet state will form when the molecule absorbs radiation. The nonradiative spin-forbidden transitions (e.g., S→T or T→S) are named intersystem crossing (ISC). It normally is very fast (<10 ns) among higher excited states (e.g., S*→T*), but may take from microseconds up to days to decay from the lowest triplet excited state to a singlet ground state (e.g., $T^* \rightarrow S_0$). The long-lived triplet state is a key point to provide enough time (above microsecond scale) for molecular interactions and allow chemical reactions to occur. By comparison the lifetime of the excited singlet state is measured in a few nanoseconds or less, which is insufficient to allow chemical reactions to occur.

Internal conversion (IC) transitions are also nonradiative but differ from intersystem crossing because electron spin remains the same (i.e., spin-allowed transition). The spin-allowed nonradiative transitions, as in IC or within vibrational modes of a particular electronic state, are exceedingly fast (usually in subnanosecond scale). Radiative transitions without spin inversion (e.g., S*→S or T*→T) are usually very fast events (<10 ns) typified by the lifetime of fluorescent photon emission. As said before, the triplet transition to the ground-state singlet takes longer to occur, and the lifetime of phosphorescence emission is generally more than 1,000 times longer than fluorescence. It is important to remark that since nonradiative transitions may dissipate some of the energy initially absorbed, photons emitted by fluorescence or phosphorescence have lower energy (longer wavelength) than the absorbed photon.

Excited molecules can alternatively be deactivated, or quenched, by donating charges or energy to a nearby ground-state molecule. Because of spin and symmetry

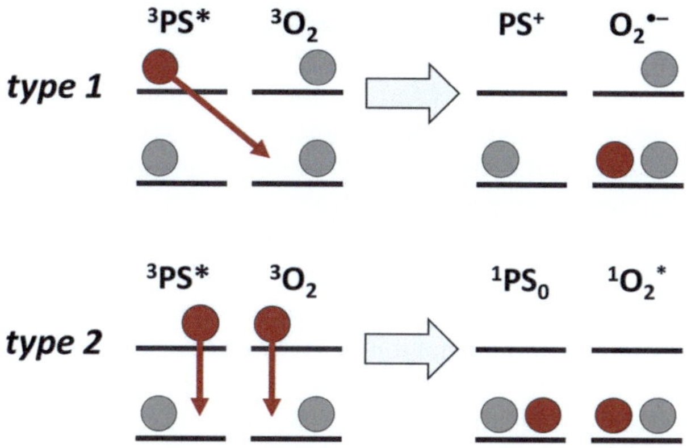

Fig. 2.7 Illustration of electronic transitions of photodynamic reactions. In type 1 reactions, excited triplet-state PS ($^3PS^*$) donates one electron to triplet oxygen ($^3O_2$) to produce superoxide radical anion ($O_2\bullet^-$). Type 2 reactions produce singlet oxygen ($^1O_2^*$) via energy donation from $^3PS^*$ to $^3O_2$ through a coulombic mechanism of electronic transition, also named as Förster or static quenching mechanism. Differently from type 1 mechanism, a dipole-dipole interaction can occur without the requirement of collision between donor and acceptor molecules and can occur in distances up to 5 nm

selection rules, i.e., triplet-triplet interactions are spin-allowed while triplet-singlet interactions are spin-forbidden, the excited triplet PS may preferentially interact with other molecules also in triplet configuration. Curiously, the vast majority of molecules found in nature are singlets in the ground state (S_0). There only a very few exceptions that are stable as triplets in their ground state, and the only one of any practical significance is molecular dioxygen (**O_2**). Also, ground-state O_2 is a diradical molecule by definition, i.e., it has two electrons alone occupying two different orbitals in the same molecule (Fig. 2.7). Being stable as a diradical triplet allows O_2 to chemically or physically interact with excited triplet molecules (e.g., photosensitizers) either via one or two electron(s) transitions. Therefore, molecular interactions between excited-state PS and ground-state O_2 is a remarkably favored process. Photodynamic reactions rely principally on such intermolecular events.

When an excited triplet-state PS ($^3PS^*$) meets triplet ground-state oxygen ($^3O_2$), two main reaction types may occur: *type 1*, one electron transfer to oxygen generating "superoxide radical anion" ($O_2\bullet^-$) and the $PS\bullet^+$ radical cation, or *type 2*, energy donation to oxygen generating excited state "singlet oxygen" ($^1O_2^*$) and ground-state PS. As a matter of fact, other secondary photochemical reactions may have some influence upon photodynamic therapy effects, depending on the photosensitizer employed. The excited PS may also donate H^+ ions to the solvent, be cleaved to become a toxic ground-state compound, or even directly interact with organic substrates to induce their release of free radicals. However, using specific chemical inhibitors (e.g., superoxide dismutase, sodium azide, dimethylthiourea, mannitol, etc.), it was demonstrated that for the best performing photosensitizers, these are only side reactions that have a minor influence upon cell killing caused by PDT [5]. Therefore, we will focus on quenching mechanisms through charge and especially energy donation, mediated by O_2, as illustrated by Figs. 2.6 and 2.7.

In type 1 reactions, $^3PS^*$ donates one electron to ground-state O_2 and yield $O_2\bullet^-$ (Figs. 2.6 and 2.7). This product is mildly reactive with biomolecules but can be reduced to hydrogen peroxide (H_2O_2), a non-radical and less reactive specie. In biological systems this reaction is constitutively catalyzed by enzymes from the superoxide dismutase (SOD) family, which occur broadly in nature [6, 7]. Hydrogen peroxide can be further reduced through reactions catalyzed by catalase (CAT) enzymes and produce $H_2O + O_2$. On the other hand, H_2O_2 may gain a second electron by transfer from the excited-state triplet PS to give the most reactive oxygen species: the hydroxyl radical (OH•). This one-electron reduction may also occur via the intermediacy of transition metal ions such as Cu^+ or Fe^{2+} (Fenton reaction).

Singlet oxygen ($^1O_2^*$) is formed as product of type 2 reactions. Even though referred as secondary reactions due to its later discovery, $^1O_2^*$ is the main ROS produced by most commercially available PS. The exact singlet oxygen production efficiency and yield is directly dependent to the photosensitizer structure, the local oxygen concentration, and which solvent they are diluted in [8, 9]. To produce singlet oxygen, the energy donation from $^3PS^*$ to $^3O_2$ takes place through a coulombic mechanism of electronic transition, also named as Förster or static quenching mechanism. In this event, $^3PS^*$ donates its deactivation energy to $^3O_2$ producing $^1O_2^*$ and $^1PS_0$ (Figs. 2.6 and 2.7). The electric dipole transition nature of coulombic (or Förster) mechanism allows $^3PS^*$ to efficiently induce the formation of $^1O_2^*$ at distances over 5 nm [10]. Due to its generally high redox potential, $^1O_2^*$ is a powerful oxygenating agent inducing cycloaddition reactions, but may also be reduced to $O_2\bullet^-$ and produce typically type 1 ROS (Fig. 2.6). Conversely $O_2\bullet^-$ can sometimes act as a reducing agent by donating the unpaired electron to a substrate, and when this happens the product of this reaction is $^1O_2^*$ (not $^3O_2$) [10]. Therefore, type 1 and type 2 pathways are not "set in stone" and crossover between different ROS can occur. The main factor influencing this crossover is the redox potential of the PS (in its ground state and its triplet state) and the redox microenvironment of the PDT reaction. Since $^1O_2^*$ is an excited-state molecule, it can also spontaneously decay to ground state via phosphorescence. The lifetime of $^1O_2^*$ in pure water is on the order of 3 μs; this permits a diffusion radius of nearly 100 nm. In the intracellular environment, the diffusion radius is much reduced because there are several targets that can react with $^1O_2^*$ [11]. Therefore, in cells $^1O_2^*$ mostly reacts with biomolecules present in the surrounding of its formation, highlighting the importance of the site of accumulation of the PS. As we will discuss in Chap. 4, $^1O_2^*$ has a ubiquitous reaction pattern with biomolecules such as aromatic and sulfur-containing amino acids, unsaturated lipids, steroids, and nucleotides [12]. Intense molecular damage caused by $^1O_2^*$ can lead to cell death via apoptosis, autophagy, or necrosis.

Depending on their electronic configuration and redox potential, each photosensitizer molecule may preferentially undergo either a type 1 or type 2 reaction, although both reaction pathways usually compete along with self-deactivation processes (Fig. 2.4). Photosensitizers that preferentially undergo type 1 photodynamic reactions are more susceptible to cellular antioxidant defense since there are specific detoxifying enzymes for ROS such as superoxide and hydrogen peroxide. Constitutive overexpression of superoxide dismutase, catalase, peroxiredoxin, and glutathiones or accumulation of manganese ions can represent effective protection against oxidation by hydrogen peroxide or superoxide and hydroxyl radicals. All these mentioned features can be sufficient to impose challenges for PDT to treat tumors and microbial strains resistant to traditional chemotherapy and radiotherapy.

Differently from type 1 ROS, there are no enzymes that are able to efficiently inactivate $^1O_2^*$, and nearly all cellular defenses rely on finite stocks of scavenger and quencher molecules that quench ROS by directly interacting with them (see Chap. 4 and 5). Since during PDT each PS molecule can produce more than 10,000 singlet oxygen molecules per second before it is destroyed by photobleaching, photosensitized cells are simply not equipped with enough antioxidant capacity to tolerate such intense attack for longer than a few minutes. In fact, there are only a few possible tolerance or resistance mechanisms to PDT ever reported, such as production of melanin granules and sequestration of PS inside them, or pumping the PS out of the cell by multidrug efflux pumps. These mechanisms of tolerance, and how to overcome them, will be posteriorly discussed in Chap. 5.

Acknowledgments MR Hamblin was supported by US NIH grant R01AI050875.

References

1. Cnossen I, Sanz-Forcada J, Favata F, Witasse O, Zegers T, Arnold NF. Habitat of early life: solar X-ray and UV radiation at Earth's surface 4–3.5 billion years ago. J Geophys Res. 2007;112(E2):10.
2. Finsen N. Phototherapy. London: Edward Arnold; 1901. 115 p.
3. Hamblin MR, Huang Y-Y. Handbook of photomedicine. New York: CRC Press; 2014. 854 p.
4. de Broglie L. A tentative theory of light quanta. Philos Mag. 1924;47(278):12.
5. Singh H, Bishop J, Merritt J. Singlet oxygen and ribosomes: inactivation and sites of damage. J Photochem. 1984;25(2):295–307.
6. Zelko IN, Mariani TJ, Folz RJ. Superoxide dismutase multigene family: a comparison of the CuZn-SOD (SOD1), Mn-SOD (SOD2), and EC-SOD (SOD3) gene structures, evolution, and expression. Free Radic Biol Med. 2002;33(3):337–49.
7. Fridovich I. Superoxide dismutases. Annu Rev Biochem. 1975;44:147–59.
8. Redmond RW, Gamlin JN. A compilation of singlet oxygen yields from biologically relevant molecules. Photochem Photobiol. 1999;70(4):391–475.
9. Wilkinson F, Helman WP, Ross AB. Rate constants for the decay and reactions of the lowest electronically excited singlet state of molecular oxygen in solution. An expanded and revised compilation. J Phys Chem Ref Data. 1995;24(2):663–77.
10. Kearns DR. Physical and chemical properties of singlet molecular oxygen. Chem Rev. 1971;71(4):32.
11. Kuimova MK, Yahioglu G, Ogilby PR. Singlet oxygen in a cell: spatially dependent lifetimes and quenching rate constants. J Am Chem Soc. 2009;131(1):332–40.
12. Bland J. Biochemical effects of excited state molecular oxygen. J Chem Educ. 1976;53(5):5.
13. Dixon JM, Taniguchi M, Lindsey JS. Photochem CAD 2: a refined program with accompanying spectral databases for photochemical calculations. Photochem Photobiol. 2005;81(1):212–3.

Chapter 3
Photosensitizers

Michael Richard Hamblin and Caetano Padial Sabino

Abstract Photodynamic therapy (PDT) was discovered over 100 years ago when it was observed that certain dyes could kill microorganisms when exposed to light in the presence of oxygen. Since those early days, PDT has mainly been developed as a cancer therapy with regulatory approvals and clinical trials steadily accumulating for different types of cancer and different photosensitizer structures. A very important milestone for PDT was the introduction of 5-aminolevulinic acid (ALA), which functions as a prodrug to induce endogenous porphyrin biosynthesis that acts as an endogenous photosensitizer produced by our cells. PDT with ALA and its derivatives have become mainstays of the clinical dermatologist's practice covering everything from skin cancer, premalignant lesions, acne, and skin rejuvenation. Another milestone in PDT development was the realization that PDT may also be used as an effective antimicrobial modality and a potential treatment for localized infections. To some extent, this means that PDT has gone full circle and returned to its roots from when it was first discovered in 1900. In this chapter we discuss, in a contextualized fashion, what are the expected characteristics of an ideal photosensitizer and

M.R. Hamblin (✉)
Wellman Center for Photomedicine, Massachusetts General Hospital,
50 Blossom Street, Bartlett Hall, Room 414, Boston, MA, USA

Department of Dermatology, Harvard Medical School, Boston, MA, USA

Harvard-MIT Division of Health Sciences and Technology, Cambridge, MA, USA
e-mail: hamblin@helix.mgh.harvard.edu

C.P. Sabino
Department of Microbiology, Institute for Biomedical Sciences, University of São Paulo,
Av. Lineu Prestes 1347, Cidade Universitária, Sao Paulo 05508-000, SP, Brazil

Department of Clinical Analysis, School of Pharmaceutical Sciences, University of São Paulo, Sao Paulo, Brazil

Center for Lasers and Applications, Nuclear and Energy Research Institute, National Commission for Nuclear Energy, Sao Paulo, SP, Brazil

Department of Medical Biophysics, Princess Margaret Cancer Institute/University of Toronto, Toronto, Canada
e-mail: caetanosabino@gmail.com

F.P. Sellera et al. (eds.), *Photodynamic Therapy in Veterinary Medicine: From Basics to Clinical Practice*, DOI 10.1007/978-3-319-45007-0_3

which are the main molecular frameworks used for development of synthetic, natural, and nanostructured photosensitizers.

3.1 Introduction

Photodynamic therapy (PDT) was discovered over 100 years ago when it was observed that certain dyes could kill microorganisms when exposed to light in the presence of oxygen (see Chap. 1). Since those early days, PDT has mainly been developed as a cancer therapy with regulatory approvals and clinical trials steadily accumulating for different types of cancer and different photosensitizer structures, starting with Photofrin® [1]. A very important milestone for PDT was the introduction of 5-aminolevulinic acid (ALA), which functions as a prodrug to induce endogenous porphyrin biosynthesis [2]. See Fig. 3.1 for a depiction of how ALA can induce accumulation of protoporphyrin IX (PpIX) in tissue to which it has been topically applied.

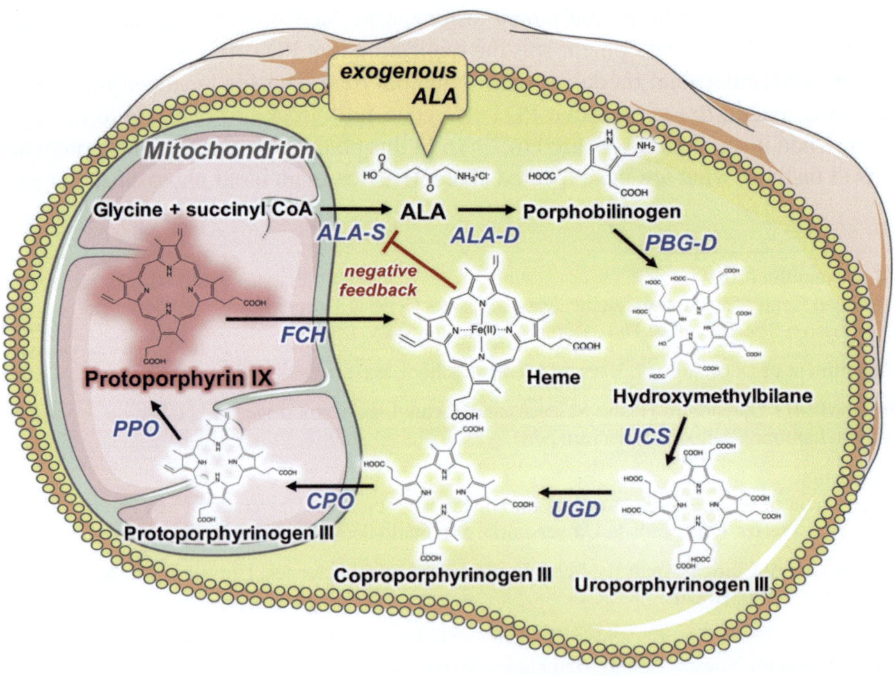

Fig. 3.1 Heme biosynthesis cycle and role of ALA-induced PPIX. *ALA-S* ALA synthase, *ALA-D* ALA dehydratase, *PBG-D* porphobilinogen deaminase, *UCS* uroporphyrinogen synthase, *UGD* uroporphyrinogen decarboxylase, *CPO* coproporphyrinogen oxidase, *PPO* protoporphyrinogen oxidase, *FCH* ferrochelatase [49]

ALA-PDT and PDT using ALA esters have become mainstays of the clinical dermatologist's practice covering everything from skin cancer, premalignant lesions, acne, and skin rejuvenation. A second major milestone in the development of PDT was its introduction as a clinical treatment for choroidal neovascularization secondary to age-related macular degeneration. The PS called Visudyne (benzoporphyrin derivative) was injected IV, and soon afterward red light was shone into the back of the eye as a way to destroy proliferating blood vessels in the retina [3]. The third milestone in PDT development was the realization that PDT may also be used as an effective antimicrobial modality and a potential treatment for localized infections [4]. To some extent, this means that PDT has gone full circle and returned to its roots from when it was first discovered in 1900. The antimicrobial applications have been enthusiastically embraced by dentists, who treat periodontitis, peri-implantitis, and endodontics [5]. The fourth milestone in PDT development (which unfortunately has not been widely accepted as yet) is the realization that PDT can stimulate an adaptive and/or innate immune response against both tumors [6] and also against pathogens [7].

3.2 Important Features of Photosensitizers

The characteristics of the ideal PS can be summarized as follows. PS should have low levels of dark toxicity to both humans and animals and a low likelihood of adverse pharmacological effects upon administration such as hypotension (decreased blood pressure) or allergic reactions. PS should absorb light in the red or far-red wavelength regions in order for the tissue-damaging effect to reach as deep as possible. It is known that both absorbance and scattering of light are minimized at longer wavelengths that penetrate the tissue deeper. Absorption bands at shorter wavelengths have less tissue penetration and are more likely to lead to skin photosensitivity (the power in sunlight drops off at $\lambda > 600$ nm). Absorption bands at too high wavelengths (>800 nm) mean that the photons will not have sufficient energy to promote PS molecules to excited triplet state to allow energy transfer to the ground state oxygen molecule to excite it to the singlet state. They should have relatively high absorption bands (>20,000 $M^{-1}cm^{-1}$) to minimize the dose of PS and light needed to achieve the desired effect. Synthesis of the PS should be relatively easy and the starting materials readily available to make large-scale production feasible. The PS should be a pure compound with a constant composition and a stable shelf life and be ideally water soluble or soluble in a biocompatible drug delivery vehicle. It should not aggregate unduly in biological environments as this reduces its photochemical efficiency. The pharmacokinetic elimination from the patient should be rapid, i.e., less than 1 day to avoid the necessity for posttreatment protection from light exposure due to prolonged skin photosensitivity. A short interval between injection and illumination is desirable to facilitate outpatient treatment that is both patient-friendly and cost-effective. Pain on treatment is undesirable, as PDT procedures do not usually require anesthesia or heavy sedation.

Although high PDT activity is generally thought to be a good thing, it is possible to have excessively powerful PS that is sometimes considered to be "unforgiving." With limitations in the effectiveness of both PS and light dosimetry, highly active PS may easily permit treatment overdosage when the surrounding normal tissue is damaged as well as the target tumor forming extensive necrotic areas. It is at present uncertain whether it is better to have a PS "tailored" to a specific indication, and to have families or portfolios of PS designed for various specific diseases or patient types, or alternatively to seek a single PS that works against most diseases. For cancer treatment it has been thought that an ideal PS should selectively accumulate in tumors after intravenous injection. Although the exact mechanisms for this "tumor-localizing effect" are not completely understood, some PS can achieve a 5:1 or higher accumulation in tumors compared to the surrounding normal tissue. Lastly, a desirable feature might be to have an inbuilt method of monitoring PS dosimetry, localization, and following response to treatment by measuring fluorescence and its loss by photobleaching in real time.

Many of the early attempts to kill microorganisms with PDT employed the same photosensitizers that were used for PDT of cancer. However, it was later realized that these structures were not optimal. Because phenothiazinium dyes (such as methylene blue) that have an intrinsic cationic charge were able to photoinactivate many classes of microorganism, it was concluded that the presence of cationic charges was crucial for broad-spectrum antimicrobial effects [8]. Although neutral and anionic PS are able to kill Gram-positive bacteria, in order to kill Gram-negative bacteria, one needs positive charges on the PS to bind and/or penetrate through the outer membrane barrier composed with large amounts of negatively charged lipopolysaccharides.

3.3 Main PS Structures

3.3.1 Tetrapyrroles

The class of tetrapyrrole PS contains (by a considerable margin) the largest number of individual compounds that have been explored for PDT both in the laboratory and also approved and tested clinically. The four most commonly explored backbone structures are porphyrins, chlorins, bacteriochlorins, and phthalocyanines. Reduction of one double bond (green arrow) in a porphyrin leads to a red shift and increase in absorption intensity of the Q-bands (e.g., 600 nm → 660 nm) in the chlorin (name derived from green chlorophyll), while reduction of two double bonds (purple arrows) leads to a further red shift (660 nm → 780 nm) and a intensity increase in the peak of the bacteriochlorin (name derived from the pigment in purple photosynthetic bacteria). The more conjugated nature of the phthalocyanine macrocycle (four additional aromatic rings) leads to a very intense absorption band at 700 nm.

Hematoporphyrin derivative (which later became known as Photofrin® and Photogem®) was approved in 1990 [9] and remains today as the most widely used PS in clinical PDT. As mentioned above, the PpIX production is induced by exogenously administered ALA, a therapeutic strategy widely used by dermatologists [10]. A later pharmacologic strategy uses methylated ALA molecules (M-ALA) to facilitate deeper prodrug diffusion through the skin using topical administration. Other porphyrin derivatives such as hematoporphyrin monomethyl ether (Hemoporfin), 2,4-di(1-methoxyethyl)-deuteroporphyrin-IX (Dimegin), and sinoporphyrin sodium [11] have been tested for PDT of cancer and a wide range of cationic porphyrins for antimicrobial PDT.

Chlorins represent a class of PS containing several examples, which have advanced as far as clinical trials. In addition to mTHPC and BPD mentioned in Table 3.1, 2-[1-hexyloxyethyl]-2-devinyl pyropheophorbide-a (HPPH), mono-N-aspartyl chlorin (e6) (Npe6) or talaporfin sodium (LS11), photoditazine, and tin(IV) etiopurpurin (SnEt2) are all classified as chlorins. Bacteriochlorins are not as stable as porphyrins and chlorins, and their use has not been as widely studied despite their more advantageous absorption peaks in the infrared region. Besides soluble TOOKAD® (Table 3.1), another compound, LUZ11 (a non-metallated fluorinated bacteriochlorin sulfonamide), has been tested in clinical trials for cancer.

3.3.2 *Synthetic Dyes*

The dye industry developed a very large number of conjugated heterocyclic compounds with high absorption bands in the visible region of the spectrum (Table 3.2). The vast majority of these dyes were found (or later designed) to be photostable so that they could be used for dyeing fabrics and clothing. However, a few of these dyes that were not photostable but were found to have an appreciable quantum yield to the triplet state that meant they could be used as PS for PDT. There remain some concerns about these synthetic dye compounds on whether they may be darker toxic compared to the tetrapyrrole compounds, which are derived from natural molecules. Most dyes were designed to be water soluble (so clothing could be dyed) that makes them suitable for antimicrobial PDT applications. Moreover, many dyes are cationic and this again encourages their use against a broad spectrum of bacteria (including Gram-negative species). Interestingly, veterinarians already use triarylmethane dyes in some countries to treat local infections and decontaminate fish tank water due to their intrinsic microbicidal activity, even though most people disregard the great antimicrobial activity enhancement that can be provided by photodynamic reactions promoted by the same dyes. In recognition to this potential, we strongly recommend those users to expose the stained water or lesions to either sunlight or artificial light sources.

3.3.3 *Other Structures*

To illustrate the diversity of different structures that have been investigated as PS in PDT, we will describe some innovative chemical structures (Table 3.3) that have not (yet) progressed as far as those shown in Tables 3.1 and 3.2.

Table 3.1 Examples of porphyrins, chlorins, bacteriochlorins, and phthalocyanines most used against cancer

Class	Name	Structure	λmax (nm)	Ref
Porphyrin	Photofrin hematoporphyrin derivative (HpD)	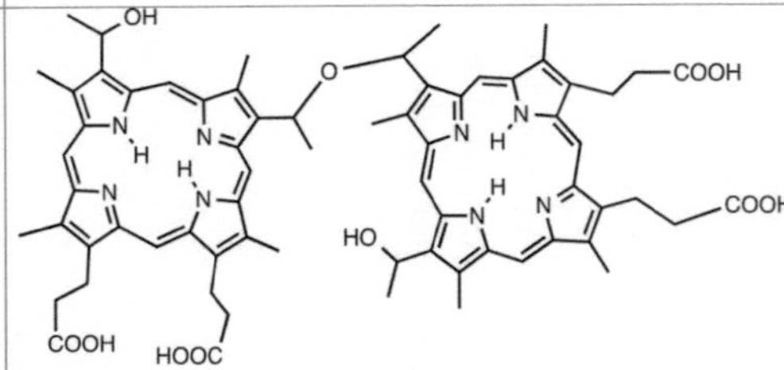	630	[12]
Chlorin	Foscan m-tetrahydroxyphenylchlorin (mTHPC)	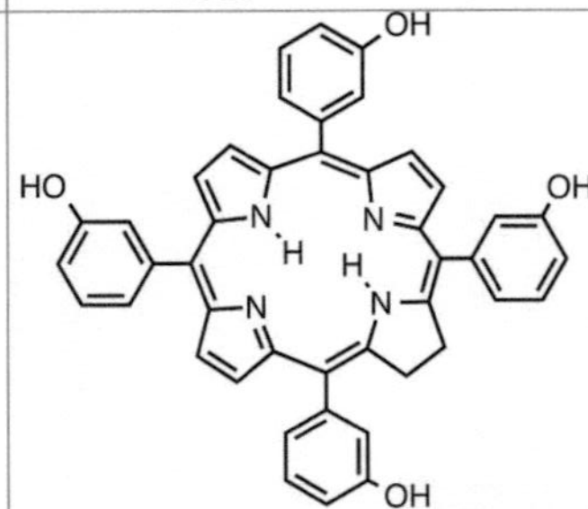	652	[13]
Chlorin	Visudyne or verteporfin Benzoporphyrin derivative (BPD)	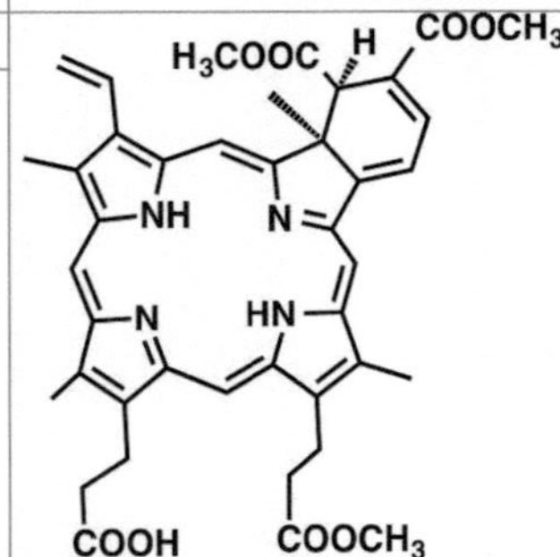	690	[14]

Bacteriochlorin	TOOKAD® (WST-11)	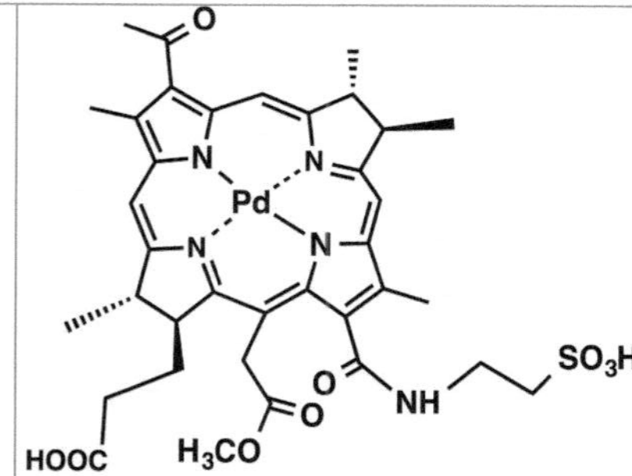	753	[15]
Phthalocyanine (Pc)	Silicon phthalocyanine (Pc4)	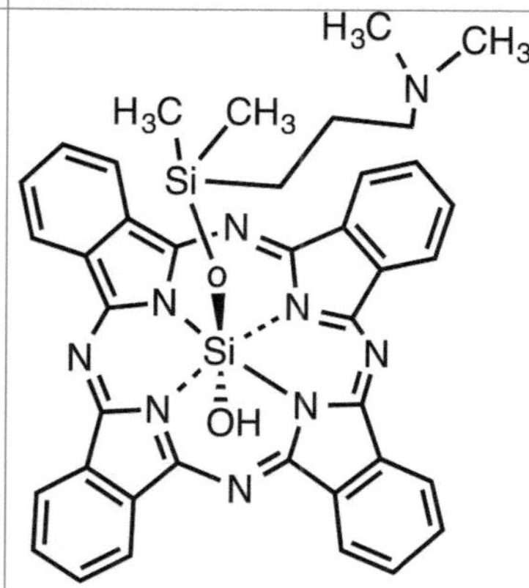	675	[16]

Table 3.2 Examples of synthetic dyes used as PS

Class	Name	Structure	λmax (nm)	Application	Ref
Phenothiazine	Methylene blue	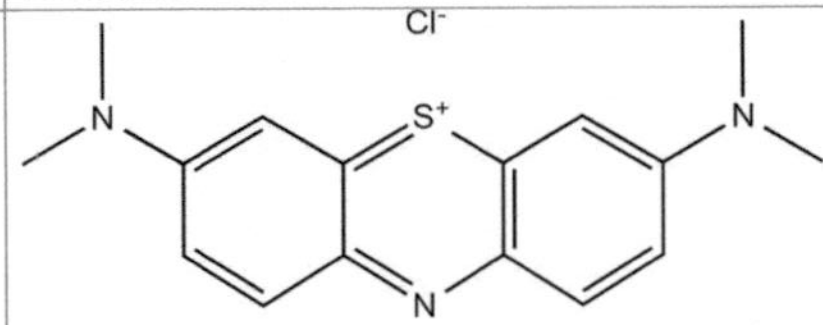	660	Cancer, antimicrobial	[17]
Benzophenothiazine	EtNBS	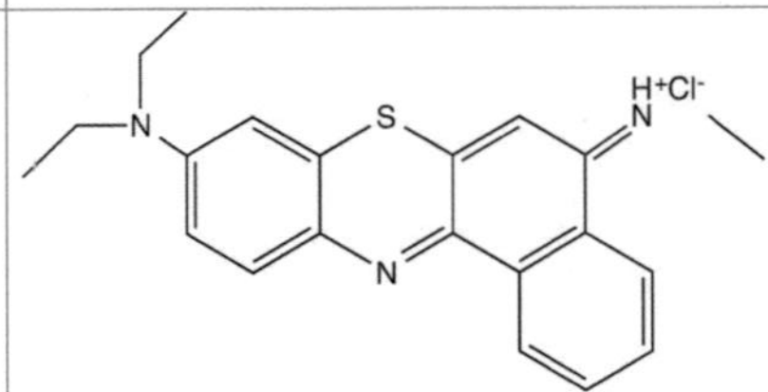	670	Cancer, antimicrobial	[18]
Halogenated xanthene	Rose Bengal	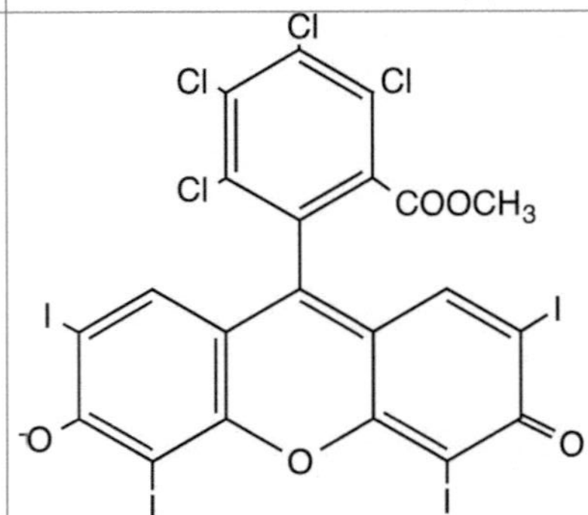	540	Cancer, antimicrobial, tissue bonding	[19]

Triarylmethane	Crystal (or gentian) violet	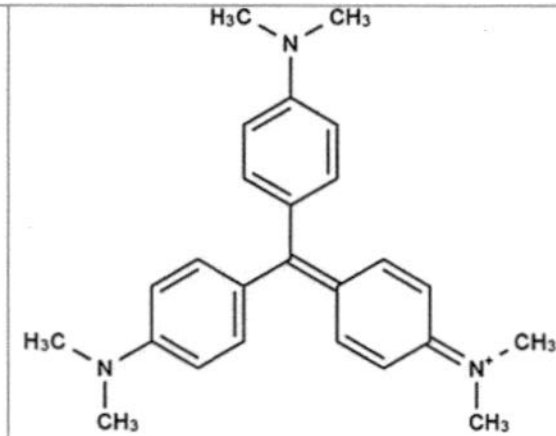	590	Antimicrobial	[20]
Triarylmethane	Malachite green	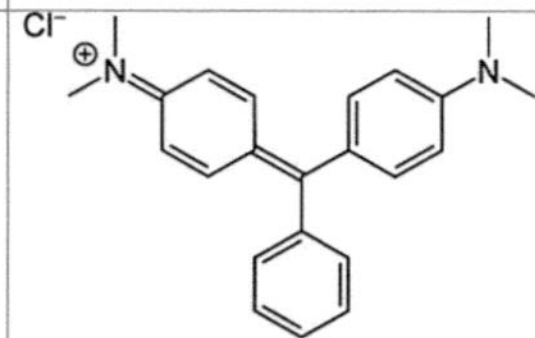	615	Antimicrobial	[21]

Table 3.3 Examples of innovative structures used as PS

Class	Name	Structure	λmax (nm)	Ref
Squaraine	ASQI	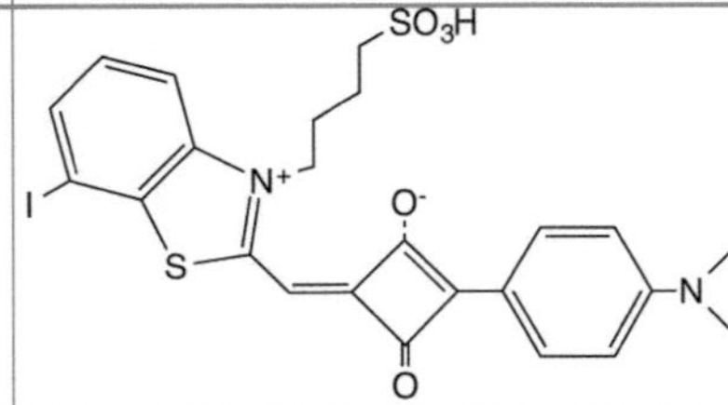	610	[22]
BODIPY	Zinc(II)-dipicolylamine diiodo-BODIPY	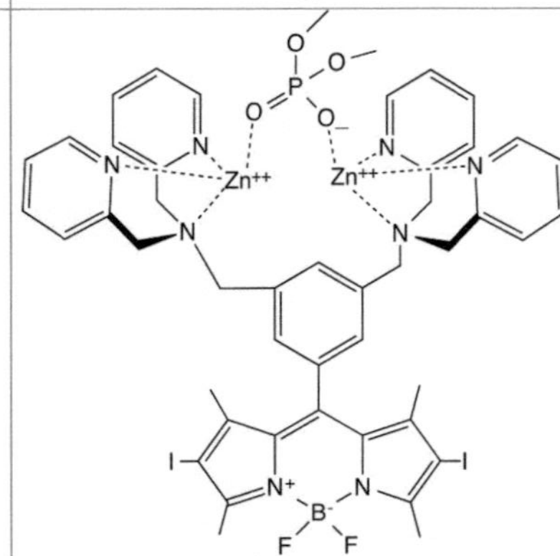	540	[23]
Phenalenone	SAPYR	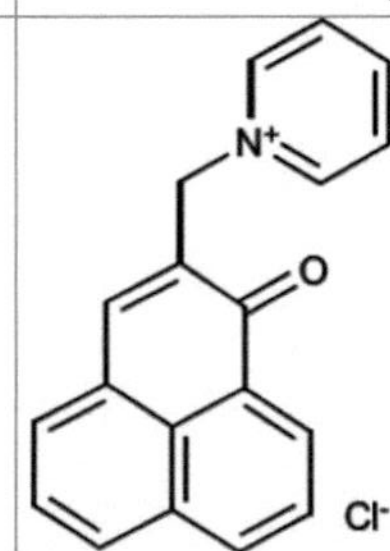	380 nm	[24]

3.3.4 PS Derived from Natural Products

In common with other areas of medicine, in PDT there is also a movement toward the use of natural products and away from synthetic drugs produced by "Big Pharma." The idea is that natural product-derived PS will have less overall toxicity and fewer side effects (Table 3.4). However, there may be a flaw in that argument in that since nearly all life-forms have evolved in sunlight, they will be unlikely to contain very powerful PS, as the consequent phototoxicity caused by sun exposure would have been selected against.

3.3.5 Inorganic Structures

Inorganic compounds have occasionally been reported to be able to carry out PDT. Two examples are given in Table 3.5. Titanium dioxide (TiO_2) presents great photostability and preferentially undergoes through type 1 photodynamic reactions (see Chap. 2). Because it is colorless but efficiently absorbs UV light, construction companies have been using TiO_2 to coat external walls of buildings forming a self-cleaning surface. As a surprising consequence, even the air quality has been reported to be better in the surroundings of TiO_2-coated buildings adding an environmental appeal for its large-scale application.

Transition metal complexes, such as $[Ru(bpy)_3]^{2+}$, are probably the most studied structures by photophysicists and photochemists over the past 40 years. Other metals such as cobalt and chromium are also often used for the same purposes. Due to spin-orbital coupling governed by the so-called heavy atom effect, intersystem crossing efficiency is extremely favored and can reach values above 99 %. This characteristic makes transition metal complexes particularly potent as PS since their singlet oxygen yield is in general the greatest possible.

3.3.6 Nanostructures

The use of nanotechnology in PDT can be broadly divided into different approaches, for example, when the nanoparticle (NP) is itself the PS (fullerenes, gold NP, silver NP, quantum dots), when the nanoparticle can absorb light to improve PDT efficiency (quantum dots, silver NP, gold NP, upconversion NP), and when the NP acts as a nano-drug delivery vehicle to improve solubility or targeting (dendrimers, liposomes, mesoporous silica NP) (see Chap. 14). Examples of these classes are shown in Table 3.6.

Table 3.4 Examples of natural product structures used as PS

Class	Name	Structure	λmax (nm)	Application	Ref
Perylenequinone (St John's wort)	Hypericin	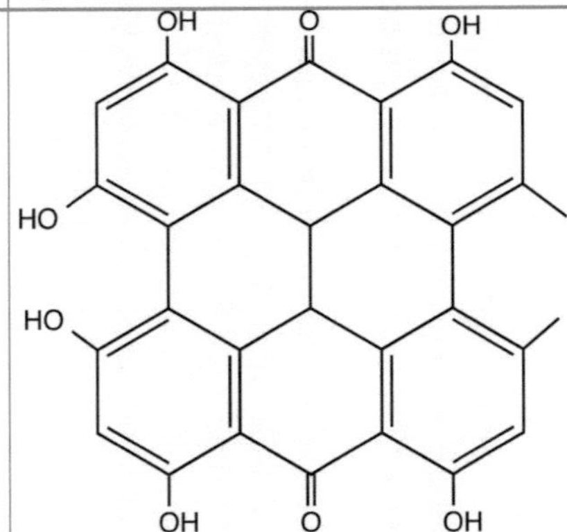	570	Cancer, antimicrobial	[25]
Flavin (vitamin B2)	Riboflavin	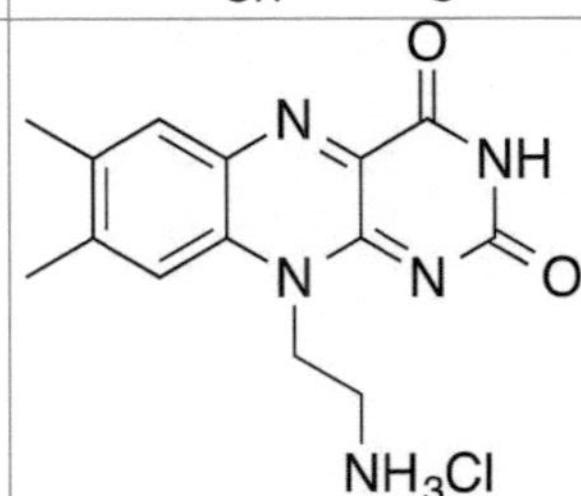	UVA or 440	Antimicrobial	[26]
Curcuminoid (turmeric spice)	SACUR-3	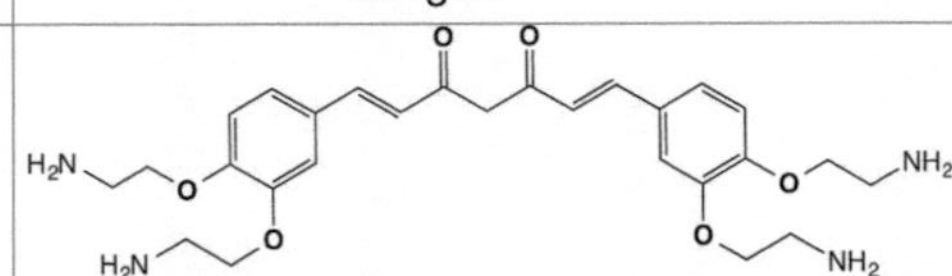	420	Antimicrobial	[27]

Table 3.5 Examples of inorganic structures used as PS

Class	Name	Structure	λmax (nm)	Application	Ref
Large band gap semiconductor	Titanium dioxide TiO_2	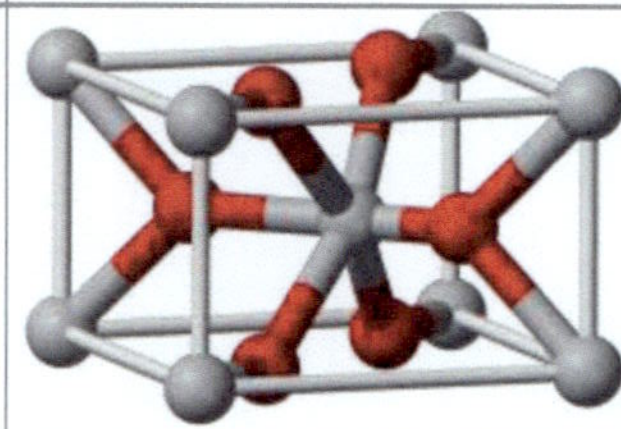	360	Antimicrobial, environmental	
Transition metal complex	Ruthenium bipyridyl Ru(bpy)	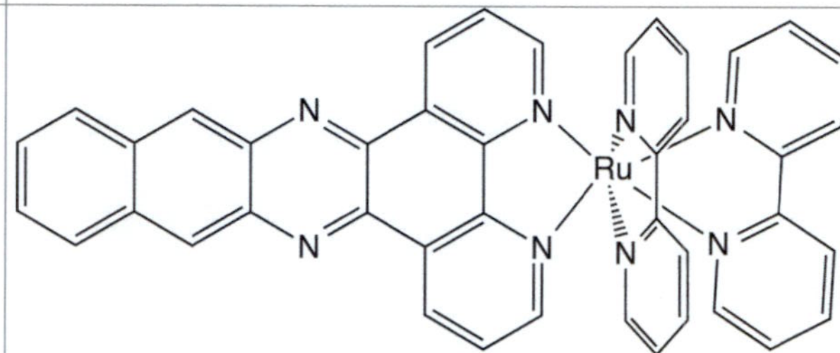	450	Antimicrobial, cancer	[28]

Table 3.6 Examples of nanostructures used in PDT

Class	PS	Structure	Goal	Application	Ref
Functionalized fullerenes	BB6	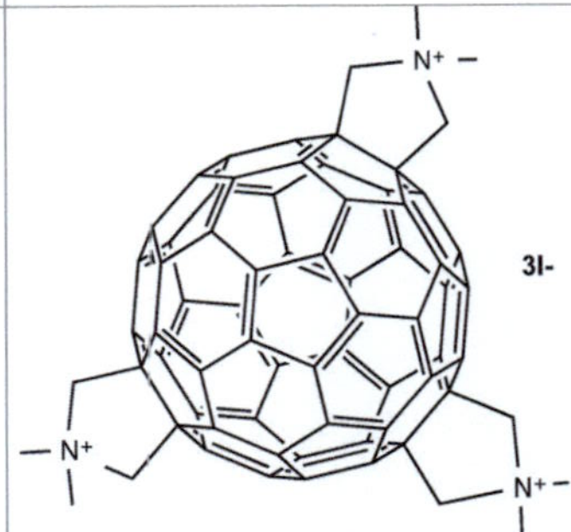	Photostable, type 1	Antimicrobial	
Gold or silver nanoparticles	Can increase activity of conventional PS	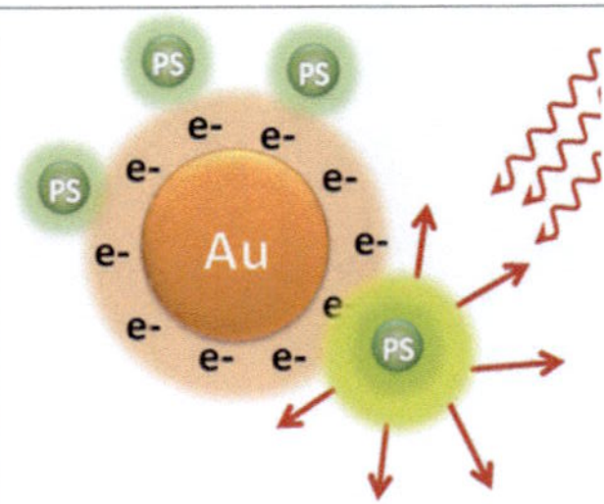	Plasmonic enhancement of PDT	Anticancer	[29, 30]
Quantum dots	Polymer QD + TPP carbon QD	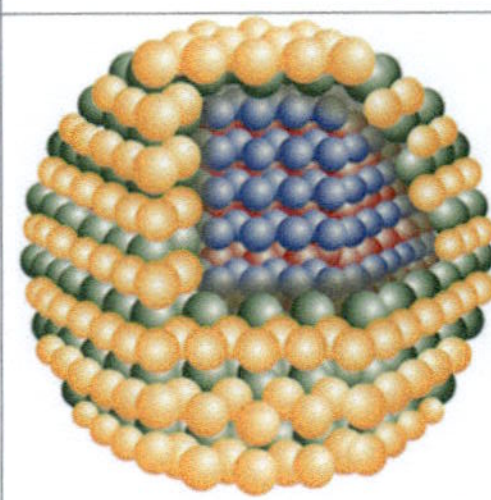	Solubilize PS, FRET (downconversion), generate ROS	Anticancer, antimicrobial	[31, 32]

Upconversion nanoparticles	Rare-earth NP/ hypocrellin A	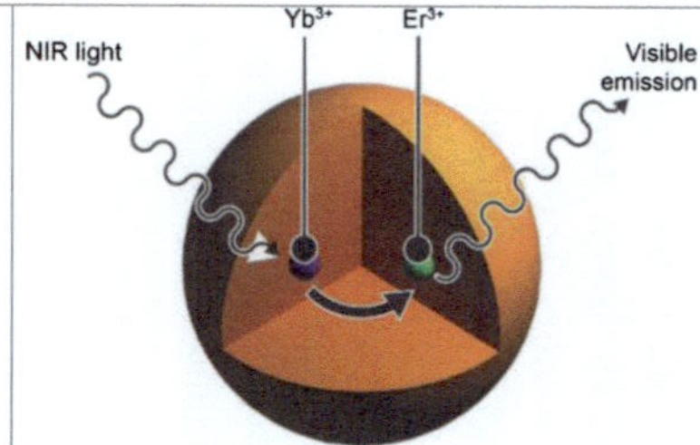	Excite with NIR light Emit blue	Cancer cells	[33]
Dendrimers	PAMAM-zinc phthalocyanine (ZnPc)	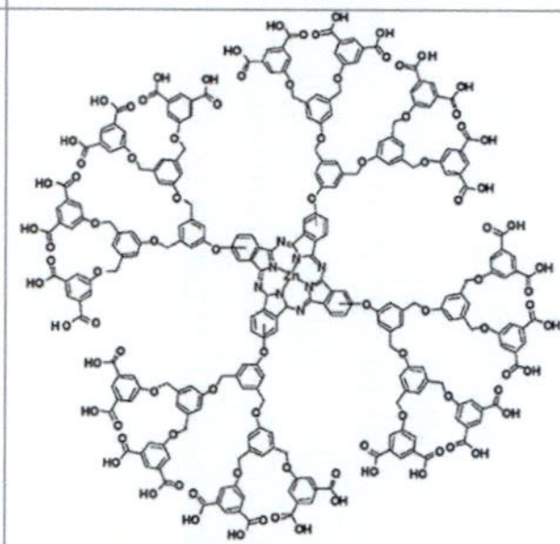	Prevent aggregation	Cancer atherosclerosis	[34, 35]
Liposomes	Foslip (liposomal mTHPC), Fospeg (PEGylated Foslip)	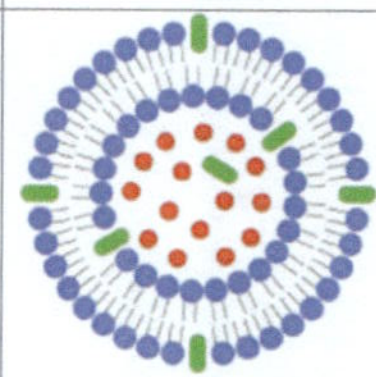	Solubilize and protect, reduce side effects	Cancer	[36]
Mesoporous silicon NP	PEI-PEG-MSN/ ZnPc	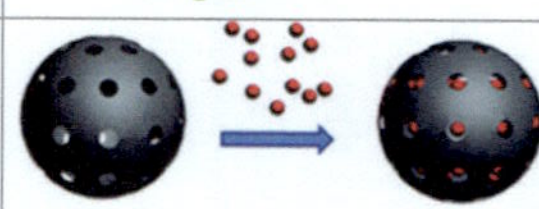	Solubilize and protect	Cancer	[37]

FRET Förster or fluorescence resonance energy transfer, *ROS* reactive oxygen species, *NIR* near infrared, *PAMAN* polyamidoamine, *PEI* polyethylenimine, *PEG* polyethylene glycol, *MSN* mesoporous silica nanoparticles

Table 3.7 Examples of targeted PS

Class	Ligand	Target	PS	Application	Ref
Monoclonal antibody	OC125	Ovarian cancer	Chlorin(e6)	Cancer	[39]
Peptide	Octreotide	Somatostatin receptor	Chlorin(e6)	Leukemia	[40]
Lipoprotein	Low-density lipoprotein (LDL)	LDL receptor	Tetra-t-butyl silicon phthalocyanine	Cancer, atherosclerosis	[41]
Serum protein	Modified albumin	Scavenger receptor	Chlorin(e6)	Atherosclerosis	[42]
Vitamin	Folic acid	Folate receptor	Pheophorbide a	Cancer	[43]
Steroid	Estradiol	Steroid receptor	Pheophorbide a	Breast cancer	[44]
Carbohydrate	Mannose	Mannose receptor	Meso-tetraphenylporphyrin	Cancer	[45]

3.3.7 PS Targeting Using Conjugates with Ligand-Receptor Recognition

Since the mechanism of preferential accumulation of various PS in tumors (or other pathological lesions) is not completely understood [38], it is rather challenging to design a PS with a high inherent ability for tumor targeting. However, there is a valid alternative approach to increase the specificity of PS for tumors. This relies on covalent conjugation of PS to various targeting ligands (either macromolecules or small molecules) that show specific molecular recognition ability with some cognate receptor that is over-expressed on tumor or other cells of interest. The advantage is that the PS can be chosen solely based on its photochemical ROS (reactive oxygen species) generation and ability to be conjugated, and the tumor-targeting component can be provided by the ligand. Examples of this ligand-receptor interaction are given in Table 3.7.

3.3.8 Nonlinear Excitation (Two-Photon, Upconversion, Second Harmonic, and Four-Wave Mixing)

The use of long wavelength light (600–1300 nm) provides deeper tissue penetration. Although some PS that absorb in the 700–800 nm range have been described, photons at wavelengths longer than 800 nm do not have enough energy to promote generation of singlet oxygen. There have been some innovative approaches developed to overcome this limitation. The most well-studied is two-photon absorption in which a very short-pulsed laser (ideally below 100 fs pulse duration) can deliver photons with such high peak power that the chances of two of them being absorbed at the same time by the PS become non-negligible [46]. The next most studied is photon upconversion by rare-earth nanoparticles that change NIR laser (often 800 or 980 nm) to visible or ultraviolet light, providing deep tissue penetration allied to site-specific generation of short wavelength light [47]. Optical upconversion can also be accomplished by second-harmonic generation in collagen, and four-wave mixing approaches, including coherent anti-Stokes Raman scattering [48].

Acknowledgments MR Hamblin was supported by US NIH grant R01AI050875.

References

1. Agostinis P, Berg K, Cengel KA, Foster TH, Girotti AW, Gollnick SO, et al. Photodynamic therapy of cancer: an update. CA Cancer J Clin. 2011;61(4):250–81.
2. Krammer B, Plaetzer K. ALA and its clinical impact, from bench to bedside. Photochem Photobiol Sci. 2008;7(3):283–9.
3. Verteporfin Roundtable P. Guidelines for using verteporfin (Visudyne) in photodynamic therapy for choroidal neovascularization due to age-related macular degeneration and other causes: update. Retina. 2005;25(2):119–34.
4. Hamblin MR, Hasan T. Photodynamic therapy: a new antimicrobial approach to infectious disease? Photochem Photobiol Sci. 2004;3(5):436–50.
5. Santin GC, Oliveira DS, Galo R, Borsatto MC, Corona SA. Antimicrobial photodynamic therapy and dental plaque: a systematic review of the literature. ScientificWorldJournal. 2014;2014:824538.
6. Castano AP, Mroz P, Hamblin MR. Photodynamic therapy and anti-tumour immunity. Nat Rev Cancer. 2006;6(7):535–45.
7. Huang YY, Tanaka M, Vecchio D, Garcia-Diaz M, Chang J, Morimoto Y, et al. Photodynamic therapy induces an immune response against a bacterial pathogen. Expert Rev Clin Immunol. 2012;8(5):479–94.
8. Merchat M, Bertolini G, Giacomini P, Villanueva A, Jori G. Meso-substituted cationic porphyrins as efficient photosensitizers of gram-positive and gram-negative bacteria. J Photochem Photobiol B. 1996;32(3):153–7.
9. Dougherty TJ, Gomer CJ, Henderson BW, Jori G, Kessel D, Korbelik M, et al. Photodynamic therapy. J Natl Cancer Inst. 1998;90(12):889–905.
10. Kostovic K, Pastar Z, Ceovic R, Mokos ZB, Buzina DS, Stanimirovic A. Photodynamic therapy in dermatology: current treatments and implications. Coll Antropol. 2012;36(4): 1477–81.
11. Wang H, Wang X, Zhang S, Wang P, Zhang K, Liu Q. Sinoporphyrin sodium, a novel sensitizer, triggers mitochondrial-dependent apoptosis in ECA-109 cells via production of reactive oxygen species. Int J Nanomedicine. 2014;9:3077–90.
12. Kessel D. Photosensitization with derivatives of haematoporphyrin. Int J Radiat Biol Relat Stud Phys Chem Med. 1986;49(6):901–7.
13. Senge MO, Brandt JC. Temoporfin (Foscan(R), 5,10,15,20-tetra(m-hydroxyphenyl)chlorin) – a second-generation photosensitizer. Photochem Photobiol. 2011;87(6):1240–96.
14. Chan WM, Lim TH, Pece A, Silva R, Yoshimura N. Verteporfin PDT for non-standard indications – a review of current literature. Graefes Arch Clin Exp Ophthalmol. 2010;248(5):613–26.
15. Moore CM, Azzouzi AR, Barret E, Villers A, Muir GH, Barber NJ, et al. Determination of optimal drug dose and light dose index to achieve minimally invasive focal ablation of localised prostate cancer using WST11-vascular-targeted photodynamic (VTP) therapy. BJU Int. 2015;116(6):888–96.

16. Anderson CY, Freye K, Tubesing KA, Li YS, Kenney ME, Mukhtar H, et al. A comparative analysis of silicon phthalocyanine photosensitizers for in vivo photodynamic therapy of RIF-1 tumors in C3H mice. Photochem Photobiol. 1998;67(3):332–6.
17. Wainwright M, Crossley KB. Methylene blue – a therapeutic dye for all seasons? J Chemother. 2002;14(5):431–43.
18. Cincotta L, Foley JW, MacEachern T, Lampros E, Cincotta AH. Novel photodynamic effects of a benzophenothiazine on two different murine sarcomas. Cancer Res. 1994;54(5): 1249–58.
19. Ali MF. Topical delivery and photodynamic evaluation of a multivesicular liposomal Rose Bengal. Lasers Med Sci. 2011;26(2):267–75.
20. Kawamoto KS,N, Shimada K, Ito K, Hirano Y, Murai S. Antibacterial effect of yellow He-Ne laser irradiation with crystal violet solution on porphyromonas gingivalis: an evaluation using experimental rat model involving subcutaneous abscess. Lasers Med Sci. 2000;15(4):5.
21. Prates RA, Yamada Jr AM, Suzuki LC, Eiko Hashimoto MC, Cai S, Gouw-Soares S, et al. Bactericidal effect of malachite green and red laser on Actinobacillus actinomycetemcomitans. J Photochem Photobiol B Biol. 2007;86(1):70–6.
22. Shafeekh KM, Soumya MS, Rahim MA, Abraham A, Das S. Synthesis and characterization of near-infrared absorbing water soluble squaraines and study of their photodynamic effects in DLA live cells. Photochem Photobiol. 2014;90(3):585–95.
23. Rice DR, Gan H, Smith BD. Bacterial imaging and photodynamic inactivation using zinc(ii)-dipicolylamine BODIPY conjugates. Photochem Photobiol Sci. 2015;14(7):1271–81.
24. Cieplik F, Spath A, Regensburger J, Gollmer A, Tabenski L, Hiller KA, et al. Photodynamic biofilm inactivation by SAPYR – an exclusive singlet oxygen photosensitizer. Free Radic Biol Med. 2013;65:477–87.
25. Theodossiou TA, Hothersall JS, De Witte PA, Pantos A, Agostinis P. The multifaceted photo-cytotoxic profile of hypericin. Mol Pharm. 2009;6(6):1775–89.
26. Maisch T, Eichner A, Spath A, Gollmer A, Konig B, Regensburger J, et al. Fast and effective photodynamic inactivation of multiresistant bacteria by cationic riboflavin derivatives. PLoS ONE. 2014;9(12):e111792.
27. Tortik N, Steinbacher P, Maisch T, Spaeth A, Plaetzer K. A comparative study on the antibacterial photodynamic efficiency of a curcumin derivative and a formulation on a porcine skin model. Photochem Photobiol Sci. 2016;15(2):187–95.
28. Lei W, Zhou Q, Jiang G, Zhang B, Wang X. Photodynamic inactivation of Escherichia coli by Ru(II) complexes. Photochem Photobiol Sci Off J Eur Photochem Assoc Eur Soc Photobiol. 2011;10(6):887–90.
29. El-Hussein A, Mfouo-Tynga I, Abdel-Harith M, Abrahamse H. Comparative study between the photodynamic ability of gold and silver nanoparticles in mediating cell death in breast and lung cancer cell lines. J Photochem Photobiol B. 2015;153:67–75.
30. Chu CK, Tu YC, Hsiao JH, Yu JH, Yu CK, Chen SY, et al. Combination of photothermal and photodynamic inactivation of cancer cells through surface plasmon resonance of a gold nanoring. Nanotechnology. 2016;27(11):115102.
31. Meziani MJ, Dong X, Zhu L, Jones LP, LeCroy GE, Yang F, et al. Visible-light-activated bactericidal functions of carbon "Quantum" dots. ACS Appl Mater Interfaces. 2016;8(17):10761–6.
32. Li S, Chang K, Sun K, Tang Y, Cui N, Wang Y, et al. Amplified singlet oxygen generation in semiconductor polymer dots for photodynamic cancer therapy. ACS Appl Mater Interfaces. 2016;8(6):3624–34.
33. Jin S, Zhou L, Gu Z, Tian G, Yan L, Ren W, et al. A new near infrared photosensitizing nanoplatform containing blue-emitting up-conversion nanoparticles and hypocrellin A for photodynamic therapy of cancer cells. Nanoscale. 2013;5(23):11910–8.
34. Spyropoulos-Antonakakis N, Sarantopoulou E, Trohopoulos PN, Stefi AL, Kollia Z, Gavriil VE, et al. Selective aggregation of PAMAM dendrimer nanocarriers and PAMAM/ZnPc nano-

drugs on human atheromatous carotid tissues: a photodynamic therapy for atherosclerosis. Nanoscale Res Lett. 2015;10:210.
35. Bastien E, Schneider R, Hackbarth S, Dumas D, Jasniewski J, Roder B, et al. PAMAM G4.5-chlorin e6 dendrimeric nanoparticles for enhanced photodynamic effects. Photochem Photobiol Sci. 2015;14(12):2203–12.
36. Senge MO. mTHPC – a drug on its way from second to third generation photosensitizer? Photodiagnosis Photodyn Ther. 2012;9(2):170–9.
37. Tu J, Wang T, Shi W, Wu G, Tian X, Wang Y, et al. Multifunctional ZnPc-loaded mesoporous silica nanoparticles for enhancement of photodynamic therapy efficacy by endolysosomal escape. Biomaterials. 2012;33(31):7903–14.
38. Hamblin MR, Newman EL. On the mechanism of the tumour-localising effect in photodynamic therapy. J Photochem Photobiol B. 1994;23(1):3–8.
39. Goff BA, Hermanto U, Rumbaugh J, Blake J, Bamberg M, Hasan T. Photoimmunotherapy and biodistribution with an OC125-chlorin immunoconjugate in an in vivo murine ovarian cancer model. Br J Cancer. 1994;70(3):474–80.
40. Kascakova S, Hofland LJ, De Bruijn HS, Ye Y, Achilefu S, van der Wansem K, et al. Somatostatin analogues for receptor targeted photodynamic therapy. PLoS ONE. 2014;9(8):e104448.
41. Demidova TN, Hamblin MR. Effect of cell-photosensitizer binding and cell density on microbial photoinactivation. Antimicrob Agents Chemother. 2005;49(6):2329–35.
42. Demidova TN, Hamblin MR. Macrophage-targeted photodynamic therapy. Int J Immunopathol Pharmacol. 2004;17(2):117–26.
43. You H, Yoon HE, Jeong PH, Ko H, Yoon JH, Kim YC. Pheophorbide-a conjugates with cancer-targeting moieties for targeted photodynamic cancer therapy. Bioorg Med Chem. 2015;23(7):1453–62.
44. El-Akra N, Noirot A, Faye JC, Souchard JP. Synthesis of estradiol-pheophorbide a conjugates: evidence of nuclear targeting, DNA damage and improved photodynamic activity in human breast cancer and vascular endothelial cells. Photochem Photobiol Sci. 2006;5(11):996–9.
45. Ballut S, Makky A, Chauvin B, Michel JP, Kasselouri A, Maillard P, et al. Tumor targeting in photodynamic therapy. From glycoconjugated photosensitizers to glycodendrimeric one. Concept, design and properties. Org Biomol Chem. 2012;10(23):4485–95.
46. Zou Q, Zhao H, Zhao Y, Fang Y, Chen D, Ren J, et al. Effective two-photon excited photodynamic therapy of xenograft tumors sensitized by water-soluble bis(arylidene)cycloalkanone photosensitizers. J Med Chem. 2015;58(20):7949–58.
47. Zhang X, Ai F, Sun T, Wang F, Zhu G. Multimodal upconversion nanoplatform with a mitochondria-targeted property for improved photodynamic therapy of cancer cells. Inorg Chem. 2016;55(8):3872–80.
48. Kachynski AV, Pliss A, Kuzmin AN, Ohulchanskyy XTY, Baev A, Qu J, et al. Photodynamic therapy by in situ nonlinear photon conversion. Nat Photonics. 2014;8:455–61.
49. Gryson O. Servier medical art France: servier. 2016. Available from: http://www.servier.com/Powerpoint-image-bank.

Chapter 4
Molecular Damage

Caetano Padial Sabino and Michael Richard Hamblin

Abstract Photodynamic therapy (PDT) rapidly produces large amounts of reactive oxygen species (ROS) to induce death of photosensitized cells. As previously described in Chap. 2, excited photosensitizer (PS) molecules can either donate electrons (type 1) or energy (type 2) to ground-state oxygen to produce superoxide radicals ($O_2\bullet^-$) or singlet oxygen ($^1O_2$). Each type of ROS has characteristic chemical reactivity and reacts with different types of chemical bonds present in biomolecules and, consequently, will lead to different types of cell damage. Once again, what determines the mechanism of cell death directly depends on both: the PS localization site within the cell and total extent of oxidative stress produced during therapy (i.e., light dosimetry and efficiency of ROS generation). To elucidate the mechanisms of photooxidative damage and the consequent biological effects, this chapter will cover the most relevant chemical reactions related to oxidative damage caused by $^1O_2$ and free radicals.

C.P. Sabino (✉)
Department of Microbiology, Institute for Biomedical Sciences, University of São Paulo, Av. Lineu Prestes 1347, Cidade Universitária, Sao Paulo 05508-000, SP, Brazil

Department of Clinical Analysis, School of Pharmaceutical Sciences, University of São Paulo, Sao Paulo, SP, Brazil

Center for Lasers and Applications, Nuclear and Energy Research Institute, National Commission for Nuclear Energy, Sao Paulo, SP, Brazil
e-mail: caetanosabino@gmail.com

M.R. Hamblin
Wellman Center for Photomedicine, Massachusetts General Hospital, 50 Blossom Street, Bartlett Hall, Room 414, Boston, MA, USA

Department of Dermatology, Harvard Medical School, Boston, MA, USA

Harvard-MIT Division of Health Sciences and Technology, Cambridge, MA, USA
e-mail: hamblin@helix.mgh.harvard.edu

F.P. Sellera et al. (eds.), *Photodynamic Therapy in Veterinary Medicine: From Basics to Clinical Practice*, DOI 10.1007/978-3-319-45007-0_4

4.1 Introduction

Photodynamic therapy (PDT) quickly generates large amounts of reactive oxygen species (ROS) to induce death of photosensitized cells. If such excessive amount of ROS could be found occurring naturally in living systems, it would indeed represent some severe pathological condition. As previously described in Chap. 2, excited photosensitizer (PS) molecules can donate either electrons (type 1) or energy (type 2) to ground-state oxygen to produce superoxide radicals ($O_2^{\bullet-}$) or singlet oxygen ($^1O_2$). Each type of ROS has characteristic chemical reactivity toward different chemical bonds present in biomolecules and, consequently, will lead to different types of cell damage. In this chapter we discuss the most relevant chemical reactions related to oxidative damage caused by photodynamic reactions to clarify the mechanisms of molecular damage caused by PDT.

4.2 Type 1 Reactions: $O_2^{\bullet-}$, H_2O_2, and $HO^{\bullet}$

Photodynamic reactions that harness energy from light to produce ROS are grouped in two types. Type 1 photodynamic reactions start via the donation of one electron from the excited PS ($^3PS^*$) to O_2 initially producing superoxide radicals ($O_2^{\bullet-}$). Superoxide can give hydrogen peroxide that can produce yet more radicals including the very reactive hydroxyl radical ($HO^{\bullet}$). Radical species are typically paramagnetic because they carry one or more unpaired electrons in valence shell orbitals. This feature makes radicals particularly reactive. They are able to react with several possible biomolecules passing the unpaired electron to another molecule also turning it into radical species or releasing another free radical. The most common radical reaction is the addition of O_2, due to its diradical nature, forming a very reactive peroxyl adduct. Hence, a single radical formed can initiate a chain reaction that can propagate to damage several biomolecules, not just one. Figure 4.1 illustrates the most relevant radical reactions for biological systems. It is noticeable that these reactions are rather unspecific and can attack lipids, sugars, proteins, and nucleic acids forming radical moieties within these biomolecules. Such radical moieties then allow biomolecules to react with each other via recombination forming cross-links between them (e.g., lipid-lipid, lipid-protein, protein-DNA, etc.). Either radical addition to biomolecules through hydroxylation and peroxidation or cross-link formation will potentially inhibit proper biomolecular function. As direct consequences, proteins are denatured and polysaccharides cleaved, nucleic acids are damaged, and membranes become less fluid and more porous and may even be disrupted. Proteins can yet be targeted for proteasomal degradation.

The radical chain reactions only cease when two radical species react with each other through dismutation or recombination. Dismutation is a classic radical reaction that can be catalyzed by enzymes from superoxide dismutase (SOD) family. It occurs when two molecules of the same radical (e.g., $O_2^{\bullet-}$) go through a redox

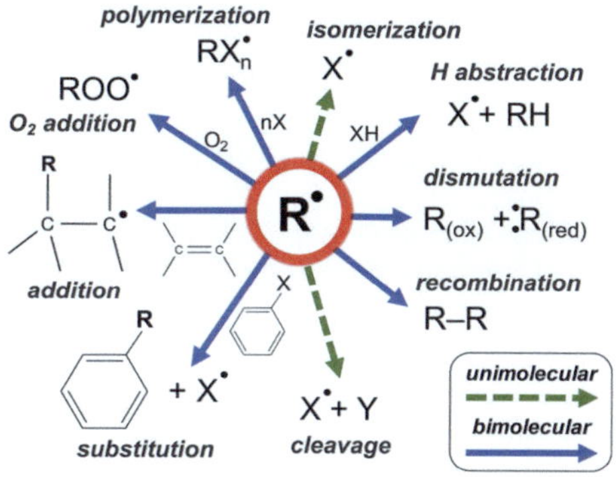

Fig. 4.1 Illustrative scheme of radical reactions most relevant for biomedical sciences. Observe that a new radical is always produced when reactions occur between radical and non-radical species. The radical chain reaction only stops when two radicals react via recombination or dismutation

reaction to produce oxidized ($^1O_2$ or $^3O_2$) and reduced (H_2O_2) products. In recombination reactions, two molecules from the same or different radicals react to form a covalently bound dimer of both radicals (Fig. 4.1).

Lipid peroxidation represents a critical source of cellular damage that often leads to necrosis. Because hydroxyl radicals are extremely reactive nonpolar species, they can rapidly abstract a hydrogen atom from saturated lipids forming lipid radicals (L•). Secondly, ground-state oxygen molecules can add to the lipid radical forming peroxyl radicals that can subsequently react with other lipids that will propagate chain reactions until recombination or dismutation reactions take place (Fig. 4.2). Since cell membrane consists of double layer of phospholipids, closely interacting with each other and with membrane proteins, radical chain reactions can spread through the membrane causing serious damages that ultimately result in necrosis via membrane rupture [1]. To minimize the extent of this type of damage, nearly all cells in nature accumulate antioxidant molecules and enzymes in membranes preventing the propagation of radical reactions [2, 3].

Special attention should be paid to nucleic acid (DNA and RNA) damage caused by HO• formed in type 1 reactions due to the diversity of damage that can be possibly produced. On the other hand, $O_2^{\bullet-}$ rarely reacts with nucleic acids due to the repulsive interaction between two negatively charged species, and although H_2O_2 is neutral, it is unable to react with most biomolecules at appreciable rates. The initial step of radical reactions in this context is most commonly associated with hydrogen abstraction from sugar moieties or addition to double bonds in nucleotides. Either pathway will lead to radical formation within the nucleic acid molecule. This might again initiate a chain reaction that will amplify the damage through the nucleic acid molecule itself or with other neighboring molecules. Consequently, HO• presents a versatile capacity to damage nucleic acid that can produce several possible damage patterns, e.g., generation of abasic sites, strand scission, DNA-protein cross-links, and inter- or intra-strand formation of nucleotide cross-links. All these damage patterns are able to impair the replication, transcription, and translation processes. Most oxidative damage patterns can be repaired by specialized DNA repair enzymes, even though formation of cross-links is often refractory to repair systems [4] (Fig. 4.3).

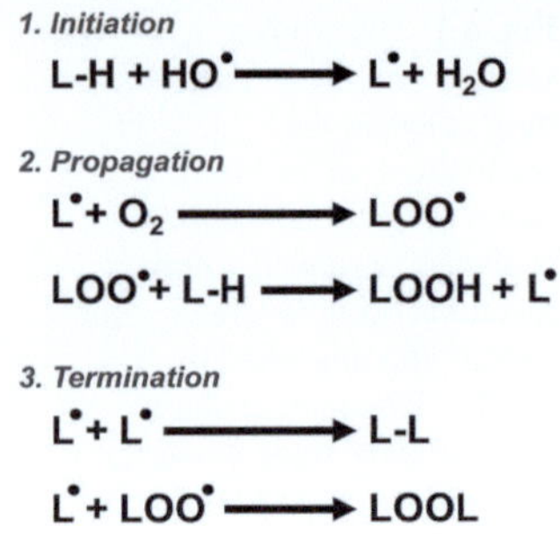

Fig. 4.2 Radical chain reactions involved with lipid peroxidation. Once chain reactions are initiated, (*1*) oxidative damage can propagate (*2*) to several other molecules until it is terminated (*3*) by recombination or dismutation reactions

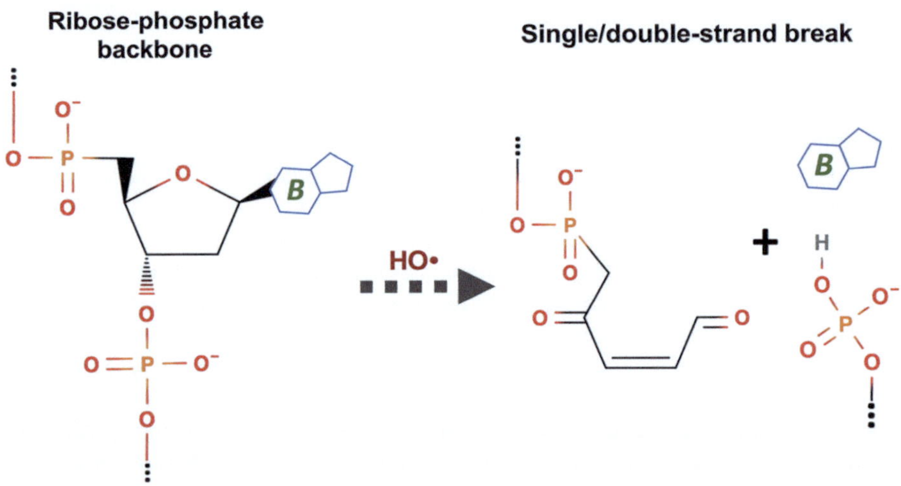

Fig. 4.3 Hydroxyl radicals (HO•) can react with ribose-phosphate backbone of nucleic acids and induce series of reactions that cause strand breaks and release of nucleobases (*B*)

Recent scientific reports, however, do not consider DNA damage as the top cause of cell death upon exposure to intense oxidative stress anymore [5]. Based on observations and studies carried out with bacteria resistant to ionizing radiation, protein damage has actually been determined to be the main cause of cell death induced by ROS. Ionizing radiation such as gamma- and X-rays have enough energy to cause radiolysis of water releasing hydroxyl radicals together with other reactive species (e.g., H_2O + gamma photon → HO• + H•). What researchers observed is that DNA damage level was very similar among radiation-resistant and susceptible bacteria. However, radiation-resistant bacteria have developed mechanisms to prevent protein damage to keep the cell viable while it repairs DNA damaged by radical reactions. All amino acids, including modified amino acids, are susceptible to damage caused by radical reactions. Protein function is highly associated to its structural conformation. Protein conformation is the result of the chemical characteristics of each individual amino acid in the polymeric chain. Of course, some amino acids are more important than others for protein structure and are active sites of enzymes; however, the oxidation of any amino acid in the chain can potentially alter protein conformation and consequently its function. In the case of membrane proteins, note that transmembrane domains are constructed with chains of hydrophobic amino acids. If these amino acids are oxidized, they become more hydrophilic and may detach from cell membrane to be solvated by water either in the cytosol or the extracellular media [6].

4.3 Type 2 Reactions: Singlet Oxygen ($^1O_2$*)

When molecular oxygen is excited to its singlet state, the spin restriction that reduces reaction rates with molecules that are not radicals is abolished. Consequently, singlet oxygen has an empty low-energy valence orbital that makes it a rather powerful oxidizing agent compared to its ground-state triplet counterpart. Singlet oxygen interacts with molecules via two basic mechanisms: physical interactions with quenchers or chemical reactions with biomolecules and scavengers. In the first mechanism, singlet oxygen transfers its excitation energy to an acceptor molecule (quencher) promoting it to its own excited state, while the oxygen returns to ground state. The excited quencher molecule can be cleaved forming new products, or else it can react with biomolecules and cause damage. Antioxidant molecules, such as carotenoids, can dissipate the energy they received from interacting with singlet oxygen in the form of heat, preventing any sort of oxidative damage to the conjugated double-bond structure. Most carotenoids are very hydrophobic molecules and exclusively localize in cell membranes where one of their major roles is to act as an antioxidant quencher [7].

Singlet oxygen mainly reacts with molecules containing thiol groups or unsaturated double bounds. Hence, it displays a more restricted pattern of reactivity as compared to free radicals produced by type 1 reactions, which react rather indiscriminately. Carbon and nitrogen atoms form double bounds sharing two electrons with opposite spin orientation. Note that ground-state triplet oxygen cannot interact with those two paired electrons because both low-energy orbitals are already occupied by one unpaired electron (see Figs. 2.6 and 2.7 in Chap. 2). Since singlet oxygen has an empty low-energy orbital, it has no restrictions on its ability to react with double bounds [7, 8].

Singlet oxygen undergoes cycloaddition reactions with compounds containing conjugated double bonds (i.e., two double bonds separated by a single bond, e.g., polyunsaturated lipids and aromatic rings) yielding peroxides (ROOR, Fig. 4.4). The oxidized molecules may undergo further reactions giving end products with hydroperoxyl (ROOH), hydroxyl (ROH), and carbonyl groups (RC=O). If only one double bond is present, singlet oxygen can induce ene reactions (Fig. 4.4). In this case, the double bond changes position, while oxygen adds to the molecule abstracting one hydrogen atom to form a hydroperoxide at the chain terminus.

Additionally, dioxetanes may be generated upon reaction with singlet oxygen when there is an electron donor atom attached to the carbon double bond. Dioxetanes are rather unstable molecules and mostly act as intermediates to yield decomposed

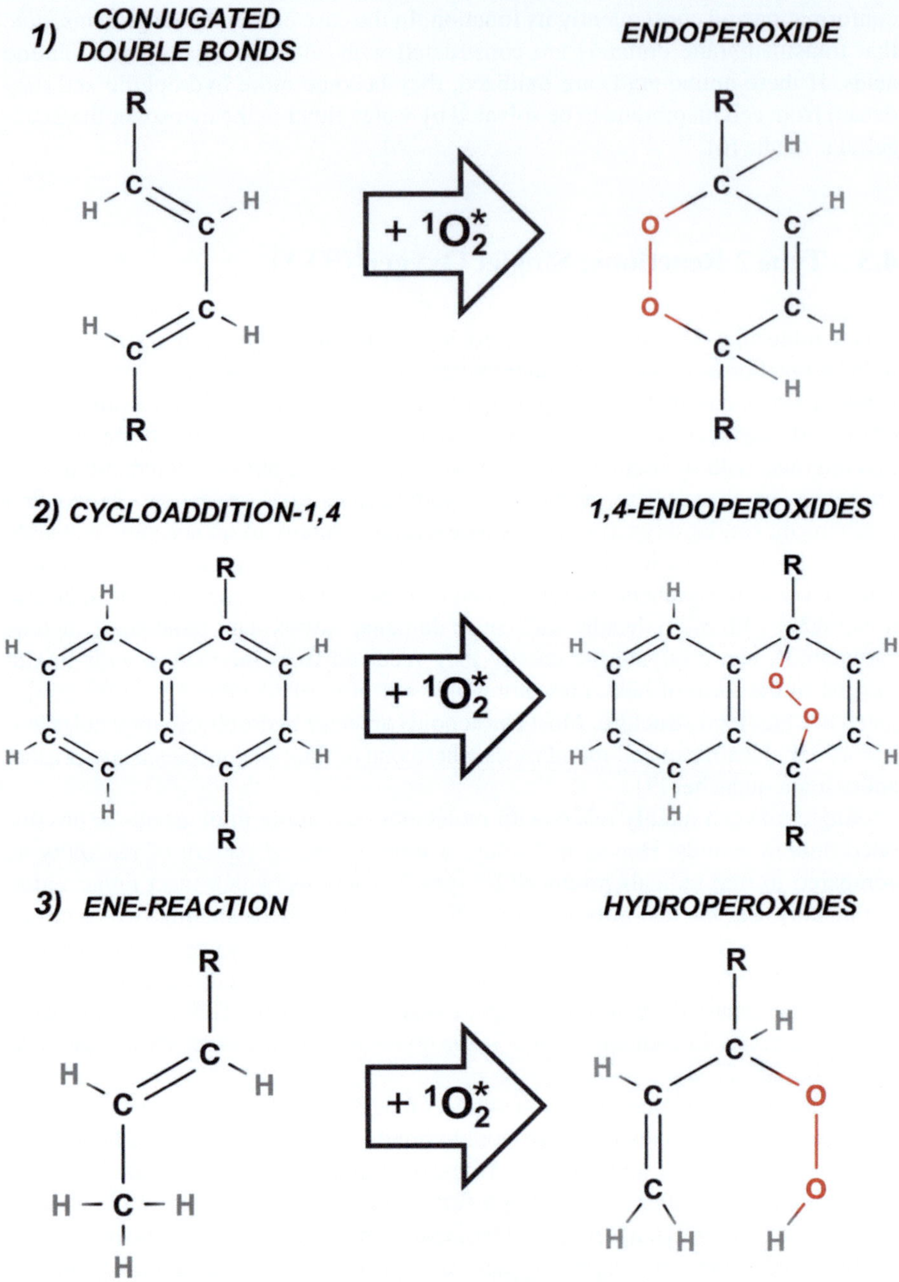

Fig. 4.4 Singlet oxygen ($^1O_2^*$) often leads to cycloaddition reactions with conjugated double bonds forming products with endoperoxide groups (*1* and *2*). Ene reactions can also take place if only one double bound is available to react with $^1O_2^*$. In this case, one hydrogen atom is abstracted and a hydroperoxide group is formed (*3*)

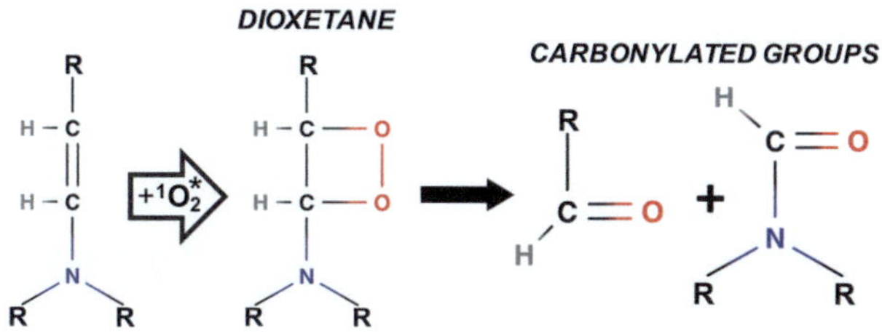

Fig. 4.5 Singlet oxygen-mediated production of dioxetane and carbonyl species. If only one double bound is present, singlet oxygen dioxetanes may be generated upon reaction with singlet oxygen when there is an electron donor atom neighboring double bounds between carbons. Dioxetanes are rather unstable molecules and mostly act as intermediates to yield decomposed products with carbonyl groups

products containing carbonyl groups (Fig. 4.5). The carbonylated products are generally stable, and in case of intracellular proteins, they can be recognized by cellular machinery that routes them for proteasomal degradation. Physicians and researchers often measure the amount of carbonylated proteins in a subject as a measure of oxidative stress [6, 9].

Unsaturated lipids such as cholesterol and essential fatty acids (e.g., omega-3 and omega-6 monounsaturated) can also react with singlet oxygen (Fig. 4.6). These lipids play a pivotal role to maintain membrane integrity and are required to produce several cytokines and hormones. Singlet oxygen reacts with double bounds of unsaturated lipids via ene reactions, hence, forming lipid peroxides. Similar to hydrogen peroxide (H_2O_2), transition metals such as Fe^{2+} and Cu^{+} can donate an electron to lipid peroxides producing free radicals that can also initiate radical chain reactions and spread the oxidative damage through the membrane. Note that this effect can potentially cause membrane rupture and is much more pronounced in mitochondrial and bacterial membranes because they harbor several proteins that contain reduced transition metal ions in their active sites, such as the enzymes involved with electron transport chain. For this reason, lipophilic photosensitizers that preferentially undergo type 2 photodynamic reactions often produce appreciable amounts of lipid peroxidation products in the cell membrane, leading to increased cell permeability but not necessarily leading to membrane rupture [1]. Also, photosensitizers that mainly produce singlet oxygen can efficiently cause mitochondrial membrane damage and lead to apoptosis.

Carbohydrates lack any double bounds between carbon atoms, and therefore they poorly react with singlet oxygen. This characteristic limitation in the reactivity of singlet oxygen with sugars differs sharply from the ability of free radicals to react with sugars and to depolymerize polysaccharides [8, 10]. Hence, singlet oxygen is more selective in its damaging ability since it cannot cleave polysaccharide chains making up microbial cell walls, glycogen, and glycoproteins nor easily break DNA/RNA strands. On the other hand, purine nucleotides (especially guanine) can be oxidized by singlet oxygen. In this reaction, singlet oxygen binds to positions 4 and 8 of the imidazole ring forming an endoperoxide intermediate that spontaneously

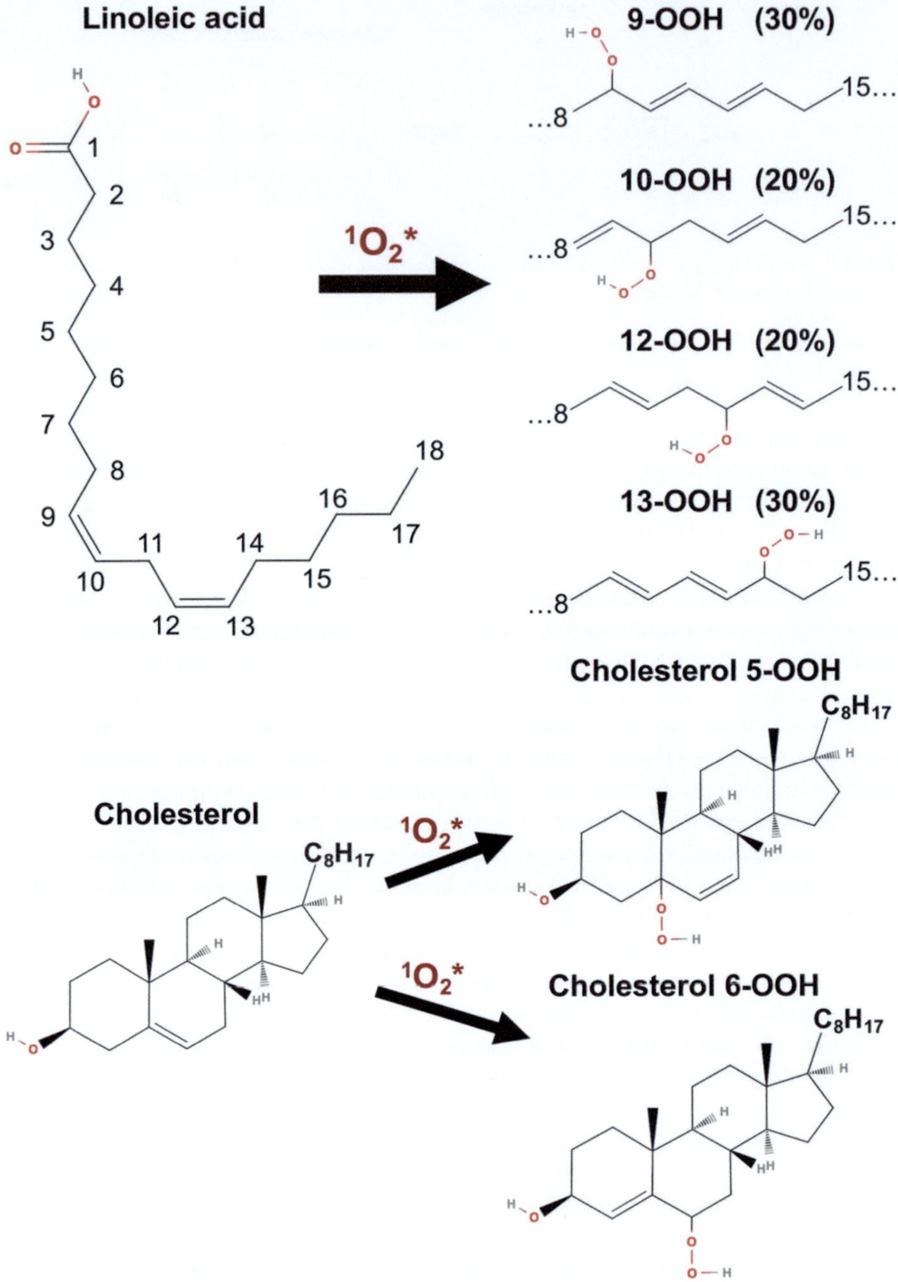

Fig. 4.6 Singlet oxygen reactions with unsaturated lipids (cholesterol and linoleic acid). Double bounds of unsaturated lipids react with singlet oxygen via ene reactions yielding lipid peroxides

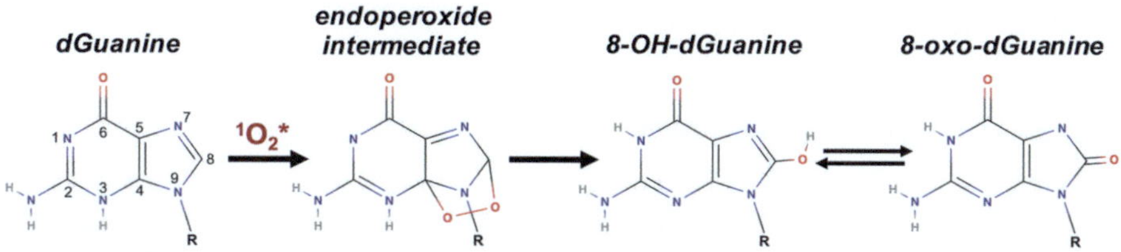

Fig. 4.7 Guanine nucleotide reactions induced by cycloaddition of singlet oxygen to positions 4 and 8 of the imidazole ring. This reaction forms an endoperoxide intermediate that spontaneously leads to oxidation of carbon 8 resulting in a hydroxyl adduct at position 8. A tautomeric equilibrium is then established between the 8-hydroxy-guanine and 8-oxo-guanine where double bonds are displaced and a hydrogen atom can bind either to the hydroxyl adduct or to the nitrogen in position 7

rearranges resulting in a hydroxyl adduct in position 8 (Fig. 4.7). This guanine species, hydroxylated in position 8, is the same as one of those produced upon reaction with HO•. A tautomeric equilibrium is then established between the 8-hydroxy-guanine and 8-oxo-guanine, where double bounds are displaced and a hydrogen atom can bind either to the hydroxyl adduct or to nitrogen in position 7. Interestingly, the oxidized base 8-oxo-d-guanine is even more vulnerable to oxidation than d-guanine and can generate diverse oxidation products such as guanidinohydantoin and oxaluric acid. Depending on which DNA polymerase operating on the damaged template strand, it may erroneously match another guanine or an adenine nucleotide to the damaged base causing G→C or G→T transversions, respectively [4]. Consequently, singlet oxygen has mutagenic potential if generated nearby chromosomal DNA, as may be the case for nucleophilic and/or cationic photosensitizers (e.g., phenothiazinium salts and acridine dyes, ruthenium complexes, etc.). It is estimated that under physiological conditions, thousands of purine bases are oxidized to 8-oxo species within our genome every day. DNA repair systems (e.g., base or nucleotide excision repairs) can correct this natural damage and minimize the mutation potential in healthy cells. However, if the level of oxidative stress surpasses the cell's capacity to repair its DNA, chromosomes may accumulate point mutations that can ultimately lead to cell death [11].

Amino acids whether free or conjugated in smaller peptide structures or in large chains within proteins can be considered as the one of the main targets of oxidation produced by singlet oxygen [6]. Proteins are present in every cell compartment and directly or indirectly participate in every functions exerted by cells. Even peptide toxins and prions, i.e., the simplest form of infectious agents that are exclusively composed of proteins, can be readily destroyed by reactions with singlet oxygen. Even when proteins and peptides suffer only low to mild damage by ROS, their function can be impaired, and they may be targeted for proteolytic degradation. However, if proteins are heavily damaged and/or aggregated, they can be refractory to proteolysis and tend to form intracellular aggregates causing further toxicity [6].

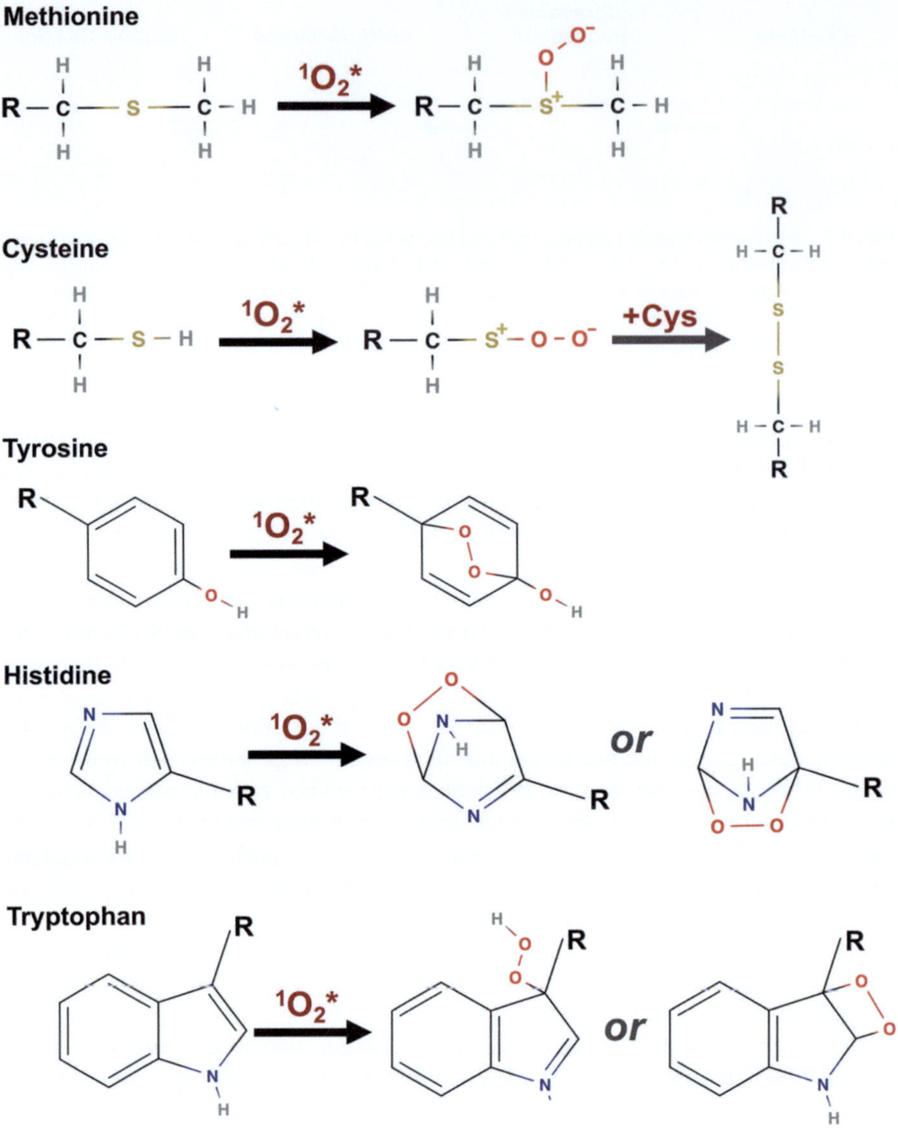

Fig. 4.8 Main singlet oxygen reactions with amino acids. Sulfur containing amino acids form peroxidic anionic species conjugated to the positively charged sulfur atom. The peroxyl adduct in cysteine residues can form a disulfide bound with other cysteine residues. For amino acids containing rings in their structure, singlet oxygen makes cycloaddition reactions producing endoperoxide intermediates that can form hydroperoxides or open the ring to yield diverse products

Among the various naturally occurring amino acids, singlet oxygen preferentially reacts with the most electron-rich residues, such as those containing aromatic rings and sulfur atoms (thiols). Cysteine and methionine react with singlet oxygen forming peroxidic anionic species conjugated to the positively charged sulfur atom

(Fig. 4.8). The peroxyl adduct in cysteine residues can receive an additional electron from transition metal donors (e.g., Cu^{+} and Fe^{2+}) and release a free radical that can react with other cysteine residues forming protein cross-links by disulfide bonds. Cysteine peroxides can yet be further oxidized yielding thiosulfinates and oxidized sulfur acids that do not induce disulfide bonds. The exact yield of each different species depends on conditions of the surrounding environment, such as concentration of other thiols or rate of ROS generation. Methionine peroxides, however, do not form disulfide bonds but mostly yield sulfoxides. Disulfide bonds and other protein-protein cross-links produce important deformations in three-dimensional structure that potentially inhibit their function. The type of damage caused by disulfide bonds is reversible and can be repaired by glutaredoxin or thioredoxin enzymes. All other types of amino acid modification due to oxidative damage are irreversible and often tag the protein to ubiquitin-independent proteasomal degradation. However, if proteins are subjected to exceedingly intense oxidative damage, they tend to form several cross-links that often produce large protein aggregates. Such aggregates are mostly refractory to proteasomal degradation and accumulate in the cytoplasm introducing further toxicity to cells.

For amino acids containing aromatic rings in their structure, such as tyrosine, histidine, and tryptophan, singlet oxygen undergoes cycloaddition reactions producing endoperoxide intermediates that can open the ring to produce diverse hydroperoxides among several other possible products [6] (Fig. 4.8). Alternatively, nitrogen atoms can donate electrons allowing the cleavage of rings and formation of carbonyl groups via dioxetane-like reactions. Since reaction kinetics between singlet oxygen and these cyclic amino acid residues are very favorable, formation of amino acid residues with hydroperoxide and carbonyl adducts is very efficient [9]. The oxidation patterns of aromatic amino acids are irreversible and lead to protein denaturation, degradation, and, potentially, aggregation.

Acknowledgments Mr. Hamblin was supported by US NIH grant R01AI050875.

References

1. Halliwell B, Chirico S. Lipid peroxidation: its mechanism, measurement, and significance. Am J Clin Nutr. 1993;57(5 Suppl):715S–24; discussion 24S–25S.
2. Buettner GR. The pecking order of free radicals and antioxidants: lipid peroxidation, alpha-tocopherol, and ascorbate. Arch Biochem Biophys. 1993;300(2):535–43.
3. Halliwell B, Gutteridge JMC. Free radicals in biology and medicine. 5th ed. Oxford, UK: Oxford Press; 2015. 944 p.
4. Dumont E, Monari A. Understanding DNA under oxidative stress and sensitization: the role of molecular modeling. Front Chem. 2015;3:43.
5. Daly MJ, Gaidamakova EK, Matrosova VY, Vasilenko A, Zhai M, Leapman RD, et al. Protein oxidation implicated as the primary determinant of bacterial radioresistance. PLoS Biol. 2007;5(4):e92.
6. Davies MJ. Protein oxidation and peroxidation. Biochem J. 2016;473(7):805–25.
7. Kearns DR. Physical and chemical properties of singlet molecular oxygen. Chem Rev. 1971;71(4):32.

8. Bland J. Biochemical effects of excited state molecular oxygen. J Chem Educ. 1976;53(5):5.
9. Dalle-Donne I, Rossi R, Giustarini D, Milzani A, Colombo R. Protein carbonyl groups as biomarkers of oxidative stress. Clin Chim Acta. 2003;329(1–2):23–38.
10. Duan J, Kasper DL. Oxidative depolymerization of polysaccharides by reactive oxygen/nitrogen species. Glycobiology. 2011;21(4):401–9.
11. Agnez-Lima LF, Melo JT, Silva AE, Oliveira AH, Timoteo AR, Lima-Bessa KM, et al. DNA damage by singlet oxygen and cellular protective mechanisms. Mutat Res. 2012;751(1):15–28.

Chapter 5
Cellular Damage

Caetano Padial Sabino and Michael Richard Hamblin

Abstract Classical pharmacology is normally concerned with defined molecular structures that can bind to specific proteins and either inhibit or enhance the protein function to achieve some biological response with therapeutic benefit. In photodynamic therapy (PDT) context, we rarely rely on such target specificity to achieve therapeutic success. Although some recent photosensitizers have been functionalized with target-specific molecules, such as antibodies, to recognize specific cells and enhance therapy specificity, ROS produced inside the cell will damage all susceptible molecules within the diffusion radius. According to the previous chapter, both hydroxyl radicals and singlet oxygen are highly reactive toward most of the abundant biological molecules contained in cells. In this chapter we discuss how such capacity of PDT to provoke multiple sites of molecular damages in the cellular context is associated with the phototoxicity produced. Also, we discuss how cellular antioxidant and xenobiotic defenses can influence on cellular tolerance against photodynamic inactivation.

C.P. Sabino (✉)
Department of Microbiology, Institute for Biomedical Sciences, University of São Paulo, Av. Lineu Prestes 1347, Cidade Universitária, Sao Paulo 05508-000, SP, Brazil

Department of Clinical Analysis, School of Pharmaceutical Sciences, University of São Paulo, Sao Paulo, SP, Brazil

Center for Lasers and Applications, Nuclear and Energy Research Institute, National Commission for Nuclear Energy, Sao Paulo, SP, Brazil
e-mail: caetanosabino@gmail.com

M.R. Hamblin
Wellman Center for Photomedicine, Massachusetts General Hospital, 50 Blossom Street, Bartlett Hall, Room 414, Boston, MA, USA

Department of Dermatology, Harvard Medical School, Boston, MA, USA

Harvard-MIT Division of Health Sciences and Technology, Cambridge, MA, USA
e-mail: hamblin@helix.mgh.harvard.edu

F.P. Sellera et al. (eds.), *Photodynamic Therapy in Veterinary Medicine: From Basics to Clinical Practice*, DOI 10.1007/978-3-319-45007-0_5

5.1 Introduction

Since the publication of the book entitled *Pathology of Oxygen*, written by A. Autor in 1982, the systemic biological effects caused by oxidative stress, including cellular damage and signaling events, have raised enough attention to become a new "hot topic" in biological sciences [1]. Indeed, following the development and refinement of reactive oxygen species (ROS) detection techniques during the 1970s it became possible to observe increased levels of free radicals associated with several health disorders such as cancer, neurodegenerative diseases, chronic inflammation, and ischemia and reperfusion-related injuries. Even today, the exact mechanisms to explain these observations concerning the various redox imbalances associated with diverse pathologies have yet to be completely elucidated, and redox biology is still a field of intensely active research.

Current scientific thinking no longer exclusively interprets ROS as sources of cellular damage [2]. Examined through an evolutionary perspective, it is logical that ROS should also play an important role as signaling molecules to regulate the metabolism of aerobic organisms. For example, the electron transport chain of cellular aerobic metabolism loses about 1 % of all transported electrons that go on to produce superoxide radicals; moreover, the phagocytic cells of our immune system use oxidative bursts that yield large amounts of ROS to inactivate pathogens. ROS are used for the synthesis of important metabolites, such as purine nucleotides and prostaglandins, where singlet oxygen seems to be a necessary reactant. More generally, most molecules present in the external environment are in oxidized forms, while in living organisms, most molecules can only function properly when reduced. Therefore, the participation of ROS in cell signaling pathways is strictly required to regulate the respiratory metabolic function and balance the redox state of cellular components.

The advent of bioinformatics and systems biology such as genomic, proteomic, and metabolomic studies has consistently corroborated these theories [3–5]. Antioxidant enzymes and ROS scavenger synthesis pathways that are responsible for redox regulation and defense against environmental sources of ROS are extremely well conserved throughout evolution (e.g., catalase, superoxide dismutase, glutathione) [5]. Conversely, gene expression regulated by oxidation-sensitive enzymes and transcription factors (e.g., IkB, HIF-1, oxyR) has been shown to govern vital cellular functions such as production of antioxidants, stage of cell cycle, inflammatory response, proliferation, motility, apoptotic signaling, and many others [2, 6]. Hence, contemporary scientific opinion offers an alternative to the old-fashioned idea that ROS were always harmful or contests the excessive simplicity of this viewpoint.

A good example of the possible dual nature of ROS is related to the practice of physical exercise. Brief bursts of metabolic activity during aerobic exercise yield moderate and short-lived amounts of ROS that can stimulate cells to upregulate antioxidant synthesis and cell growth pathways leading to protection against future oxidative insults, while exercise simultaneously induces muscle hypertrophy and

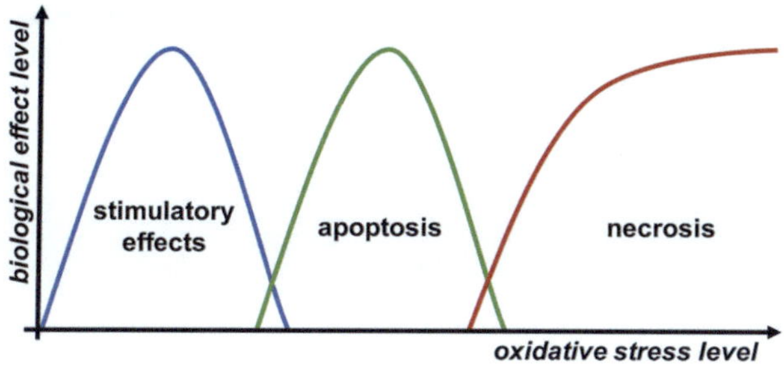

Fig. 5.1 Illustrative plot of biological effects induced by oxidative stress. In simple terms, low levels of oxidative stress stimulate cell metabolism, intermediary levels induce apoptosis signaling, and high levels can lead to direct necrosis. Autophagy is considered as an extreme survival strategy that occurs in oxidative stress levels where apoptosis and stimulatory effects overlap. Hence, autophagic cells may end up surviving or dying depending on a variety of factors

proliferation. On the other hand, if oxidative stress or physical exercise is too intense or too prolonged, damage caused by ROS may be sufficient to induce harmful effects such as inflammation, cell death, and tissue injury. Hence, the balance between beneficial and detrimental effects of ROS relies on its production site, rate, duration, and total yield (Fig. 5.1). Also, the balance between apoptotic and necrotic cell death pathways induced by ROS damage depends directly on the level of oxidative stress along with other factors such as the PS localization.

Classical pharmacology is normally concerned with defined molecular structures that can bind to specific proteins and either inhibit or enhance the protein function to achieve some biological response with therapeutic benefit. Hence, the molecular targets and mechanisms of action of most chemotherapy drugs can be precisely determined, and side effects are usually linked to lack of drug specificity. In PDT context, we rarely rely on such molecular target specificity to achieve therapeutic success. Although some recent photosensitizers have been functionalized with target-specific molecules, such as antibodies, to recognize specific cells and enhance therapy specificity, ROS produced inside the cell will damage all susceptible molecules within the diffusion radius. According to the previous chapter, both hydroxyl radicals and singlet oxygen are highly reactive toward most of the abundant biological molecules contained in cells. In this chapter we discuss how such molecular damages in the cellular context are associated with phototoxicity produced by PDT.

5.2 Ultrastructural Damage: How Does ROS Kill Cells?

Highly reactive oxygen species generated by photodynamic reactions, such as hydroxyl radicals (HO•) and singlet oxygen ($^1O_2$), can rapidly react with all molecular structures that are involved in the maintenance of proper cell physiology and viability. As described over the previous chapter, type 1 and type 2 reaction products can modify components of nucleic acids, proteins, and cell membranes inducing

Table 5.1 Reaction rate constants (k, L mol^{-1}s^{-1}) for hydroxyl radicals (HO•) and singlet oxygen ($^1O_2$) with some examples of relevant biomolecules. Proteins and DNA rate constants are based on human-derived macromolecules

Target	HO•	$^1O_2$
DNA	8×10^8	5.1×10^5
dG	7.6×10^9	5.3×10^6
dA	4×10^9	$<1\times10^5$
Albumin	7.8×10^{10}	5×10^8
Trypsin	1.6×10^{11}	8×10^9
Lysozyme	5.1×10^{10}	1.3×10^8
Histidine	1.3×10^{10}	9×10^7
Tryptophan	1.3×10^{10}	6.6×10^7
Cysteine	3.4×10^{10}	8.3×10^6
Methionine	7×10^9	8.6×10^6
Glycine	6×10^6	$<1\times10^5$
Arginine	3.5×10^9	–
Leucine	1.7×10^9	–
Linoleic acid	9×10^9	1×10^7
Glucose	1×10^{10}	1.4×10^4
Ascorbate	1×10^{10}	8.3×10^6
GSH	1.4×10^{10}	2.4×10^6

Data was obtained from literature publications [7, 8]

either reversible or irreversible damage. If sufficient damage were selectively delivered to any of those structures, it would already be enough to completely inactivate any cell. However, when dealing with highly reactive oxidants, and especially with HO•, we must expect to observe a ubiquitous pattern of widespread cellular damage. In Table 5.1 we show the reaction rate constants of HO• and $^1O_2$ with some examples of the most important biomolecules present in membranes, nucleic acids, saccharides, and proteins. Even though singlet oxygen and peroxides are more selective in the targets they react with, all biomolecular structures mentioned above are susceptible to oxidation in some specific positions.

The diffusion radius of highly reactive species tends to be very small in biological samples. Therefore, ROS react with the biomolecules closest to their production site, i.e., the region of photosensitizer accumulation. The overall reaction rate is determined by the reaction rate constant multiplied by the target concentration. Since the PS concentration varies among different compounds (Figs. 5.2a–c) and cell types, and even among cellular regions or compartments, the actual oxidative impact of PDT on each cellular biomolecule is most strongly influenced by PS concentration at the ROS production site. Hence, the amount of damage distributed to each cellular structure mostly depends on where the photosensitizer accumulates, the concentration of targets at that region, the rate constants for the reaction between oxidants and targets, the concentration of antioxidants, and the cellular capacity to repair oxidative damage. Based on all these factors, computer simulations suggest that the oxidant reactivity alone is insufficient to determine which cellular targets

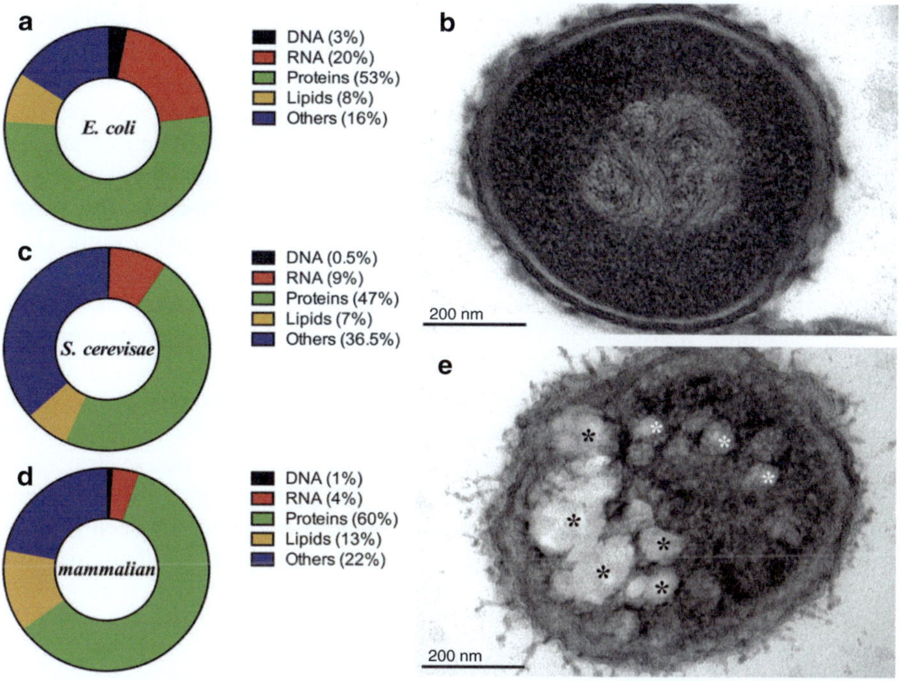

Fig. 5.2 From (**a**–**c**) we present the most abundant biomolecules present in *Escherichia coli* (**a**), *Saccharomyces cerevisiae*, (**b**) and human leukocytes (**c**) (Data from Feijo Delgado et al. [38]). Note that proteins are the major components of cells. Transmission electron microscopy images of transversal sections of a Gram-negative bacterium (*Klebsiella pneumoniae*) before (**d**) and after (**e**) being inactivated by exposure to methylene blue-mediated PDT. Clear regions indicated by asterisks (*) are large protein aggregates formed by intense oxidation imposed by PDT (Transmission electron microscopy images are from C. P. Sabino's personal collection)

are most damaged. In fact, the damage ratios done by HO• and $^1O_2$ are nearly equal among nucleic acids, fatty acids, proteins, and other structures [7].

So far, most scientific investigations have focused their attention on oxidative damage to DNA and membrane lipids, while relegating protein oxidation to secondary importance. Indeed, lipid peroxidation can rapidly kill cells by membrane rupture, and DNA oxidation can also kill cells or lead to mutations. Another point is that DNA and lipid oxidation are relatively easier to experimentally quantify and are much less complex to interpret. However, over the past years, some rather important reports have been indicating that protein oxidation may actually be the key process behind ROS-derived cell signaling and toxicity. Until recently it was believed that ionizing radiation kills cells by DNA oxidation due to reaction with HO• generated by water homolysis. However, using a model of radioresistant bacteria (*Deinococcus radiodurans*), Daly et al. demonstrated that the primary reason for cellular inactivation upon gamma irradiation was actually imposed by protein oxidation [9]. This publication brought important changes to the paradigm on mechanisms of cellular inactivation by ROS and added further importance to the determinants of protein (per) oxidation. Unfortunately, the PDT scientific community is still conducting experiments to determine whether proteins are also a primary target for cell inactivation by $^1O_2$. However, if we consider theoretical simulations, we can still expect that $^1O_2$ might also inactivate cells via protein damage, i.e., in the same way that HO• does [7]. Proteins are ultimately responsible for most cellular functions and are scattered through all compartments of all cells. Regardless of taxonomy or strain, proteins are the most abundant component of cells (Figs. 5.2a–c). In cell membranes they compose from 30 to 75 % of the total membrane dry weight [10]. Bacterial cells lack intracellular compartments or organelles. Their DNA is centrally distributed within the cell and is surrounded by large amounts of protein complexes, including transcription factors and ribosomes. In Fig. 5.2, we present transmission electron microscopy images of a Gram-negative bacterial cell before (Fig. 5.2d) and after (Fig. 5.2e) inactivation by exposure to methylene blue and red light. It seems clear that the cell membrane is not disrupted by PDT, because all the intracellular contents are still contained inside the cell. However, several clear spheres (indicated by asterisks, *) appear after irradiation. Those spheres, often referred as inclusion bodies, are large protein aggregates commonly seen in cells that were exposed to oxidative stress or intense heat. Although it is difficult to quantify, these images clearly illustrate the importance of protein oxidation during PDT.

5.3 Cell Death Pathways

Eukaryotic cells, in general, present a much more complex spatial and molecular organization when compared to prokaryotes such as bacteria. Due to the electrochemical diversity found inside each organelle of eukaryotic cells, it has been possible to employ photosensitizers with molecular structures that show a distinct tendency to accumulate in certain intracellular compartments. In contrast to what can be seen in Fig. 5.2d, e, in eukaryotic cells damage may be localized to certain organelles, while the other cellular compartments remain undamaged. Hence, ROS production can be targeted to specific structures that will trigger cell death pathways which correspond to the signaling cascades arising from specific damage to different sites (Fig. 5.3). Eukaryotic cells of metazoan animals present rather elaborate signaling pathways leading to regulate either cell death or survival. Necrosis and apoptosis are the two main types of cell death subroutines that have been classically established based on morphologic and biochemical divergences. In this classic perspective, apoptosis would be a programmed and "organized" pathway of cell death, while necrosis is classified as unregulated or "accidental" death. Even though the morphological and biochemical aspects do indeed exist, whether it is a matter of programmed or accidental, subroutine is not so simple to distinguish. Here we will describe the mechanisms based on studies of the mammalian class of organisms, because the literature still lack studies with respect to reptiles, birds, amphibians, etc.

Autophagy is often mistakenly interpreted as an independent mechanism of cell death. It actually is a "last ditch" mechanism to try to rescue severely damaged cells

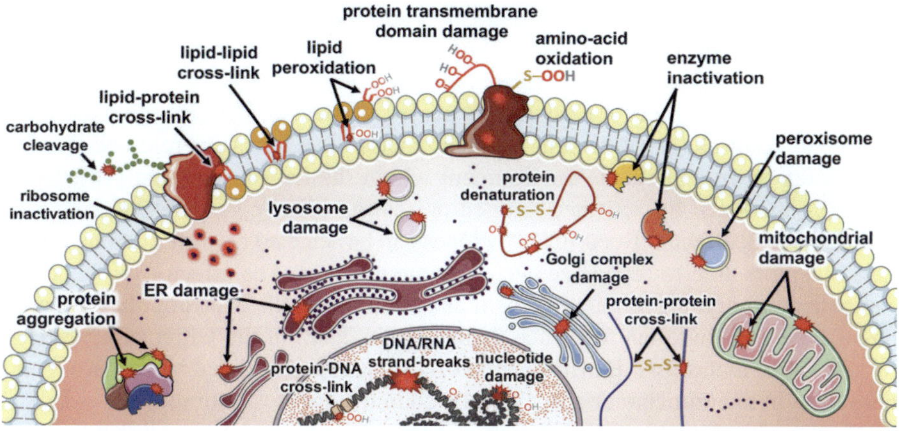

Fig. 5.3 Illustrative examples of cellular damage caused by photodynamic therapy. Note that virtually all vital structures of the cell can potentially be damaged and lead to diverse cell death pathways. As described in this chapter, the factors that determine the preferential damage site are photosensitizer localization, photodynamic reaction type, and target concentration [39]

from death. In this situation, the damaged proteins and organelles are digested in lysosomes to increase protein turnover and (by eliminating nonfunctional structures) supply amino acids to rebuild them [11, 12]. It is clear that excessive degradation of cellular structures can lead to cellular death; however, in well-controlled situations, it can indeed rescue cells from apoptosis or necrosis [13]. Autophagy can be initiated by light-activated photosensitizers that accumulate either in the cytosol, endoplasmic reticulum (ER), mitochondria, or lysosomes. After damage is detected, the autophagic signal is propagated through the cell via mTOR/AKT pathway. Interestingly, cells with a blocked apoptosis pathway either due to impaired signaling or lack of ATP may also independently initiate autophagy to promote cell death [13].

Apoptosis was initially described by pathologists from a morphological point of view. Cells undergoing apoptosis lose their original shape as they shrink and form apoptotic bodies and membrane blebs containing degraded intracellular contents. The chromatin is condensed and DNA is cleaved in regions located between nucleosomes. Apoptosis is completely regulated by genetic information, and, hence, some refer to it as "programmed cell death." Nearly all apoptotic pathways depend directly or indirectly on mitochondrial signaling via proteins from the caspase family. The intrinsic pathway can be initiated by the presence of certain ROS-damaged organelles or molecules in the cytosol that can be sensed by mitochondrial proteins. Caspase-9 signaling then promotes expression of proapoptotic genes to induce the programmed cell death routine. Alternatively, cells can undergo a caspase-independent apoptosis via release of mitochondrial proteins such as apoptosis-inducing factor (AIF), Omi/HtrA2, or endonuclease G [14, 15].

The extrinsic apoptosis pathway is originated by extracellular stimuli mediated by immune cells or activation of surface pro-death receptors. Adaptive immunity can be directed against some particular cell types (these may be some kind of immunogenic cancer or virus-infected cells), especially if characteristic epitopes are exposed to membrane surface in a non-tolerogenic environment. PDT-mediated oxidative damage can induce a trauma in treated tissues so that damage-associated molecular patterns (DAMPs) and proinflammatory cytokines are released thereby signalizing an immunogenic environment within the damaged region [16]. Upon signaling mediated by activated antigen-presenting cells (e.g., dendritic cells and macrophages), effector T-cell ($CD8^+$) clones can be selected to proliferate with specific receptors able to recognize such cognate epitopes. The effector lymphocytes can patrol through the whole organism and locate the cells exposing their specific cognate epitope and then to induce the apoptotic death of the target cells, even in distant tumor metastasis. Extrinsic apoptosis signaling is initiated by caspase-8 and converges in the mitochondria to elicit the common apoptosis subroutine.

Necrosis has long been regarded as a severe trauma-associated cell death that is passive and uncontrolled. However, it was recently described that necrotic cell death can also be regulated by ROS, caspase-8, and receptor-interacting protein 1 (RIP1) [17]. Either way, necrosis is characterized by cellular swelling, membrane rupture, and leakage of cytosolic content including organelles, proteins, nucleic acids and other DAMPs [16, 18]. These intracellular contents are highly immunogenic and potentially lead to progression of inflammation and adaptive immunity. PDT may induce necrosis if PS accumulates in cell membrane (e.g., highly lipophilic molecules or compounds that do not enter cells) or if PDT doses are exceedingly high.

5.4 Organelle Damage

5.4.1 Ribosomes

Ribosomes are very large protein-RNA complexes that are responsible for the synthesis of all proteins within our cells. Containing both protein and RNA regions, ribosomes present several oxidation labile sites that can lead to their functional inactivation [19]. This type of organelle damage is deadly to the cell since it directly impairs protein synthesis, including translation of antioxidant defense and repair systems in response to changes in cellular redox balance.

5.4.2 Mitochondria

As mentioned above, mitochondria play a pivotal role in apoptosis signaling cascades. Many photosensitizers, especially the cationic amphiphilic ones, can selectively accumulate in mitochondria and will cause rapid permeabilization and destruction after light activation. High levels of lipid peroxidation in the external mitochondrial membrane can open the membrane permeability pores that allow passage of caspase activators such as cytochrome c, AIF, and "second mitochondria-derived activators of caspases" (SMAC) such as DIABLO [14, 15].

5.4.3 Lysosomes

Lysosomes are organelles responsible for degradation of both intracellular and extracellular biomolecules that are damaged or unwanted. Their interior is maintained at an acidic pH level (pH 4.5–5), and they contain more than 50 enzymes responsible for the degradation of many biological substrates. Lysosome-targeted PDT can cause the membrane to rupture thereby releasing all the contents into the cytoplasm. PDT-damaged lysosomes also release cathepsins that can promote mitochondrial membrane permeabilization via cleavage of enzymes from the BH3-only family or can even directly activate caspases to initiate the intrinsic apoptosis pathway [20].

5.4.4 Endoplasmic Reticulum

Endoplasmic reticulum (ER) stress can lead to several different signaling pathways that may lead to a pro-survival outcome or to intrinsic apoptosis pathways and cell death. Which pathway will eventually decide the fate of the cell is determined by the level of oxidative stress that the organelle was exposed to. Low to mild levels of ROS can activate survival pathways via induction of autophagy, antioxidant production, and JNK and $p38^{MAPK}$ signaling pathways that are responsible for cell survival and even promotion of growth. On the other hand, exposure to high levels of oxidative stress can increase the permeability of the ER membrane causing release of calcium ions that elicit mitochondrial-mediated intrinsic apoptosis [21]. PDT-mediated ER stress has also been associated with activation of CCAAT-enhancer-binding proteins (C/EBPs) homologous protein (CHOP) that is also inducible by DNA damage. The CHOP transcripts induce expression of apoptosis-associated genes that define the cell's fate [22].

5.5 Antioxidant Defense and Resistance Mechanisms

The development of cellular antioxidant defense enzymes and pathways was a milestone in the history of evolution. About 1.5 billion years ago, the abundance of photosynthetic organisms in the planet started to impact the atmospheric O_2 concentration. At first, all life on the Earth thrived using variations of anaerobic metabolism, while O_2 represented only a toxic by-product of photosynthesis that had to be avoided to prevent cellular damage. Hence, O_2-rich environments on the Earth left gaps to be colonized by those life-forms which could better withstand such harsh conditions. This was the turning point for natural selection of organisms that developed antioxidant defense mechanisms to keep a balanced intracellular redox status [23]. After this evolutionary turnaround, cells were able to develop new metabolic pathways to utilize the chemical energy of O_2 to facilitate ATP production via oxidative phosphorylation with much greater efficiency than the previous anaerobic fermentation process.

Proteomic studies of organisms from all the kingdoms of life currently inhabiting our planet have revealed that there is a highly conserved group of about 300 proteins that are considered to be the minimal essential set of genes that can produce a viable organism. Out of those 300 proteins, 44 (15 %) are dedicated to the response and defense against cellular stress, and 18 (6 %) are essential proteins for maintenance of the correct intracellular redox balance and avoiding oxidative damage to macromolecules and membrane lipids [4]. Therefore, the pivotal importance of cellular mechanisms of antioxidant defenses either in physiological or pathological conditions remains unquestionable. Since PDT kills cells via oxidative damage, the cell's capacity to regulate its own redox balance and prevent oxidative stress can directly influence its sensitivity to photodynamic inactivation. For this reason, PDT protocols must be developed to ensure that oxidative stress imposed to target cells surpasses their antioxidant capacity. Following, we will describe how cellular antioxidant systems work and how PDT can overcome them.

Living organisms can utilize small molecular weight antioxidants to basically act as redox balance buffers. Scavenger or quencher compounds are easily oxidized and form nonreactive products. Physical quenchers mainly dispose the excess energy of singlet oxygen by changing to a nonreactive excited state (and ground state oxygen) that rapidly decays to the ground state dissipating energy in the form of heat. So when small-molecule antioxidants are present in micromolar to millimolar intracellular concentrations, they can minimize oxidation of important biomolecules simply by competition (i.e., they interact with ROS before other molecules do it).

Some small molecular weight antioxidants are constantly produced by animal cells (e.g., urate, glutathione, melanin), and others must be acquired in the diet (e.g., vitamin C for humans, vitamin E, carotenoids). Many antioxidants obtained from diet are derived from vegetables. Plants produce large amounts of scavengers and quenchers because chlorophyll has a finite probability to undergo photodynamic side reactions that yield ROS while carrying out normal photosynthesis. This is why plant leaves that have been overexposed to the sun can become yellowish and die. After the green chlorophyll has been oxidized to colorless compounds, the yellowish color of leaves reveals the abundance of carotenoid pigments produced by plant cells to prevent oxidation of chloroplasts and other organelles. Carotenoids such as ß-carotene and lycopene are among the most potent lipophilic antioxidants. Their molecular structures characteristically present several conjugated double bounds that can rapidly scavenge radicals or singlet oxygen. Upon radical reactions, it forms a nonreactive radical where the unpaired electron is delocalized through all conjugated double bounds. Additionally, carotenoids can scavenge singlet oxygen by cycloaddition as well as physically quench it forming a nonreactive excited state. However, carotenoids are highly lipophilic and can only accumulate in membranes, so they leave hydrophilic biomolecules relatively unprotected. This may explain why carotenoid-producing bacteria, such as *Staphylococcus aureus*, are one of the most susceptible bacterial species to photodynamic inactivation using hydrophilic photosensitizers [24].

Vitamin C (ascorbate) and E (tocopherols) are complementary small-molecule antioxidants that best perform their function when combined together. Both contain double bounds that allow efficient reaction with singlet oxygen or with radicals (and these antioxidants themselves form nonreactive radicals); however, while ascorbate is highly soluble in water, tocopherols are rather lipophilic. Tocopherols are excellent scavengers of lipid peroxyl radicals and, hence, can terminate chain reactions of lipid peroxidation. Moreover, ascorbate can then act as a reducing agent and regenerate from tocopheryl radicals. Ascorbyl radicals can then undergo dismutation reactions forming dehydroascorbate and a regenerated ascorbate or be reduced by specific enzymes (e.g., NADH-dehydroascorbate reductase family). Hence, these vitamins can be efficient antioxidants for type 1 photodynamic reactions (i.e., radical production). Tocopherol can act as a singlet oxygen quencher or a nonrecyclable scavenger, giving some degree of protection for cell membranes. On the other hand, ascorbate provides very poor cellular protection since it can only scavenge singlet oxygen at relatively low rates (see Table 5.1).

Urate (uric acid) is a product of purine degradation that accumulates in many animals that are deficient in enzymes responsible for its further decomposition to more water-soluble compounds. Birds, reptiles, Dalmatian dogs, and primates (including humans) are classic examples of urate-accumulating animals. Urate is a powerful antioxidant capable of scavenging radicals, peroxides, ozone, and singlet oxygen. Its respective oxidized radical can be recycled by ascorbate similarly to tocopheryl radicals [25]. Hence, successful PDT treatment of tumors in urate-accumulating animals may require higher doses compared to treatment of other animals although this has not yet been established in well-controlled experiments.

Reduced glutathione (GSH) is a tripeptide thiol antioxidant present in huge concentrations (1–50 mM) in nearly all eukaryotic and Gram-negative bacterial cells [3]. In eukaryotic cells, GSH is found in equal concentrations in the cytosol and nucleus, but not inside mitochondria. It is not produced in most Gram-positive bacterial species, but some species have membrane transporters that allow uptake of GSH from mammalian hosts. GSH efficiently scavenges radicals, quenches/scavenges singlet oxygen, and is an enzymatic substrate for reduction of peroxides and hydroperoxides (Fig. 5.4) [26]. In a similar manner to the cysteine residues of proteins, oxidized glutathione can form disulfide bounds with another glutathione (GSSG) or with protein thiols. Glutathione disulfides do not react with biomolecules, but can be reduced back to GSH by glutathione reductase enzymes. The cellular content of GSH is directly related to the cell's capacity to tolerate photodynamic inactivation or gamma radiation. Cells can survive longer radiation exposure times compared to their GSH-deficient counterparts. GSH provides simultaneous protection of DNA, proteins, and membrane lipids [27–29]. It has also been demonstrated that fractionated PDT doses decrease phototoxicity due to regeneration of GSH by glutathione reductases in between light doses [30].

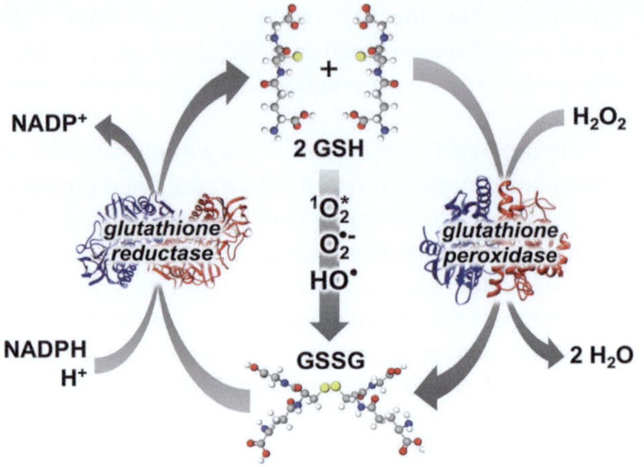

Fig. 5.4 Glutathione-dependent antioxidant defenses are broadly available in animals and microorganisms and can be highly effective to protect cells against several oxidants. The cysteine residue of reduced glutathione (GSH) can be oxidized and lead to formation of disulfide bonds with another cysteine residue of glutathione (GSSG) or proteins. GSSG can be reduced back to two GSH in a reaction catalyzed by enzymes with glutathione reductase activity, at the cost of NAPDH and H^+. Hydrogen peroxide can be reduced by two GSH molecules catalyzed by glutathione peroxidase, yielding two molecules of H_2O

In addition to small molecular weight antioxidants, cells also rely on redox-balancing enzymes that can catalyze deactivation of ROS. Among the 18 essential proteins for intracellular redox balance, the most relevant enzymes present in animal and microbial cells are superoxide dismutase (SOD), catalase (CAT), and the family of enzymes that use glutathione as a substrate. However, none of these enzymes are able to directly protect cells from singlet oxygen or hydroxyl radicals (Fig. 5.5). SOD was first described in 1969 as an abundant enzyme of vital importance present in virtually all animal tissues. SOD catalyzes the dismutation of superoxide (into hydrogen peroxide and oxygen) by several orders of magnitude, and it is indeed fundamental to protect cells from ROS generated by cell aerobic metabolism or xenobiotics. However, SOD is only able to impair photodynamic inactivation that is mediated by the few photosensitizers that preferentially undergo type 1 reactions, such as triarylmethanes (e.g., crystal or gentian violet and malachite green) and titanium dioxide semiconductors. Hence, it is best to use type 2 photosensitizers to avoid this cellular defense mechanism. The dismutation reaction produces an oxidized (O_2) and a reduced (H_2O_2) species. Therefore, superoxide dismutase detoxification must be accompanied by the decomposition of H_2O_2 to water and nontoxic products. Many enzymes such as CAT, glutathione peroxidase (GPx), and peroxiredoxin (Prx) are able to perform this task. While CAT directly catalyzes H_2O_2, dismutation to H_2O and O_2, GPx, and Prx carries out a more complex sequence of reactions. GPx catalyzes the reaction between H_2O_2 and two GSH forming H_2O and GSSG that must subsequently be regenerated to GSH by glutathione reductase. Prx has cysteine residues that can be directly oxidized by H_2O_2 forming disulfide

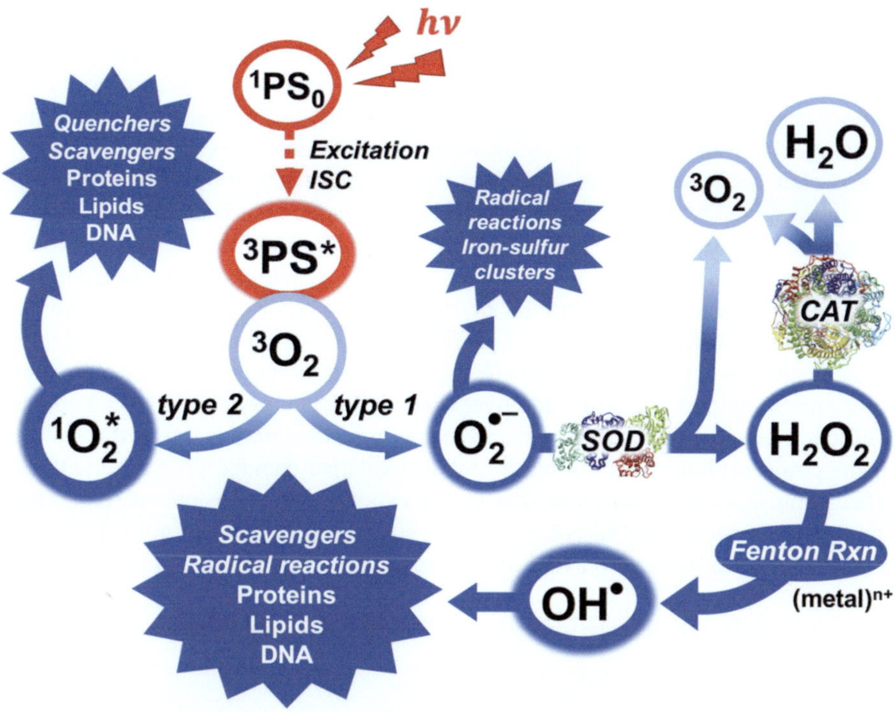

Fig. 5.5 General overview of type 1 and 2 photodynamic reactions in the cellular context. Ground state photosensitizer ($^1PS_0$) is excited by light (*hv*) to a higher energy state that decays to the long-lasting excited triplet state ($^3PS^*$) via intersystem crossing (ISC). Type 1 reaction initiates via charge transfer from $^3PS^*$ to $^3O_2$ forming superoxide radical anion ($O_2{\bullet}^-$). Superoxide can react with antioxidant scavengers, iron-sulfur clusters of enzyme active sites, or undergo dismutation reaction, catalyzed by superoxide dismutase (SOD), producing O_2 and hydrogen peroxide (H_2O_2). Hydrogen peroxide can be decomposed to H_2O and O_2 in reaction catalyzed by catalase (CAT) or other enzymes. Hydrogen peroxide can also be reduced by transition metals (e.g., Fenton reaction with Fe^{2+} or Cu^+) producing hydroxyl anions (HO^-) and radicals (HO•). Hydroxyl radicals are extremely reactive and can induce radical chain reactions in proteins, lipids, nucleic acids, and carbohydrates. Type 2 reactions produce singlet oxygen ($^1O_2^*$) via energy transfer. Singlet oxygen is also extremely reactive toward proteins, nucleic acids, and lipids. It can be quenched or scavenged by small molecular weight antioxidants, but, just like HO•, it is not impaired by any enzymes

bounds. Thioredoxin (Trx) is able to protect cells from oxidative stress by becoming oxidized itself, thus removing disulfide bounds from oxidized proteins such as Prx to restore the native protein structure. Finally, Trx needs to react with NADPH to be regenerated back to its reduced active form.

Melanin pigments constitute some of the most difficult challenges for PDT and other therapeutic modalities. In parallel to the potent antioxidant activity that makes melanotic melanomas resistant to radiotherapy and PDT, it also acts as an optical barrier for light penetration into tissues and can even sequester photosensitizer molecules due to its anionic nature. Several recent and ongoing investigations have had some degree of success trying to overcome or, at least, minimize this limitation using tissue clearing techniques and photosensitizers that absorb infrared light (e.g., bacteriochlorins or naphthalocyanines) [31–33]. Either way, melanin-producing tumors or pathogens (e.g., *Cryptococcus neoformans*) still represent an additional challenge for nearly all commercially available photosensitizers.

Last but not least, another important resistance mechanism to PDT, also commonly expressed by drug-resistant tumors and pathogens, is the efflux of photosensitizers through membrane drug transporters [34–37]. The most commonly found types belong to the family of ATP-binding cassette transporters (ABC-transporters). In many cases the coadministration of verapamil or other efflux pump inhibitors could inhibit the resistance phenotype [35–37]. Even so, this drug is not highly specific and can cause several side effects (related to inhibition of other membrane transporters) that limit its use. Moreover, since efflux pumps are in constant contact with photosensitizer substrates, photodynamic reactions may occur in the vicinity of the pump leading to its inhibition due to oxidative damage [34]. Alternatively, other photosensitizer molecules that act in extracellular environment (e.g., large conjugated molecules or vascular targeting strategies) or that are not substrate of efflux pumps can be used to avoid resistance to PDT [34].

Acknowledgments Mr. Hamblin was supported by the US NIH Grant R01AI050875.

References

1. Autor AP. Pathology of oxygen. New York: Academic; 1982. 384 p.
2. Halliwell B, Gutteridge JMC. Free radicals in biology and medicine. 5th ed. Oxford, UK: Oxford Press; 2015. 944 p.
3. Copley SD, Dhillon JK. Lateral gene transfer and parallel evolution in the history of glutathione biosynthesis genes. Genome Biol. 2002;3(5):research0025.
4. Kultz D. Molecular and evolutionary basis of the cellular stress response. Annu Rev Physiol. 2005;67:225–57.
5. Landis GN, Tower J. Superoxide dismutase evolution and life span regulation. Mech Ageing Dev. 2005;126(3):365–79.
6. Zelko IN, Mariani TJ, Folz RJ. Superoxide dismutase multigene family: a comparison of the CuZn-SOD (SOD1), Mn-SOD (SOD2), and EC-SOD (SOD3) gene structures, evolution, and expression. Free Radic Biol Med. 2002;33(3):337–49.
7. Davies MJ. The oxidative environment and protein damage. Biochim Biophys Acta. 2005;1703(2):93–109.
8. Wilkinson F, Helman WP, Ross AB. Rate constants for the decay and reactions of the lowest electronically excited singlet state of molecular oxygen in solution. An expanded and revised compilation. J Phys Chem Ref Data. 1995;24(2):663–77.
9. Daly MJ, Gaidamakova EK, Matrosova VY, Vasilenko A, Zhai M, Leapman RD, et al. Protein oxidation implicated as the primary determinant of bacterial radioresistance. PLoS Biol. 2007;5(4):e92.
10. Guidotti G. Membrane proteins. Annu Rev Biochem. 1972;41:731–52.
11. Kroemer G, Levine B. Autophagic cell death: the story of a misnomer. Nat Rev Mol Cell Biol. 2008;9(12):1004–10.
12. Shen S, Kepp O, Kroemer G. The end of autophagic cell death? Autophagy. 2012;8(1):1–3.

13. Shen S, Kepp O, Michaud M, Martins I, Minoux H, Metivier D, et al. Association and dissociation of autophagy, apoptosis and necrosis by systematic chemical study. Oncogene. 2011;30(45):4544–56.
14. Oleinick NL, Morris RL, Belichenko I. The role of apoptosis in response to photodynamic therapy: what, where, why, and how. Photochem Photobiol Sci. 2002;1(1):1–21.
15. Kessel D, Luo Y. Mitochondrial photodamage and PDT-induced apoptosis. J Photochem Photobiol B. 1998;42(2):89–95.
16. Garg AD, Krysko DV, Vandenabeele P, Agostinis P. DAMPs and PDT-mediated photo-oxidative stress: exploring the unknown. Photochem Photobiol Sci. 2011;10(5):670–80.
17. Vanlangenakker N, Vanden Berghe T, Krysko DV, Festjens N, Vandenabeele P. Molecular mechanisms and pathophysiology of necrotic cell death. Curr Mol Med. 2008;8(3):207–20.
18. Fonseca C, Dranoff G. Capitalizing on the immunogenicity of dying tumor cells. Clin Cancer Res. 2008;14(6):1603–8.
19. Singh H, Bishop J, Merritt J. Singlet oxygen and ribosomes: inactivation and sites of damage. J Photochem. 1984;25(2):295–307.
20. Stoka V, Turk B, Schendel SL, Kim TH, Cirman T, Snipas SJ, et al. Lysosomal protease pathways to apoptosis. Cleavage of bid, not pro-caspases, is the most likely route. J Biol Chem. 2001;276(5):3149–57.
21. Hubmer A, Hermann A, Uberriegler K, Krammer B. Role of calcium in photodynamically induced cell damage of human fibroblasts. Photochem Photobiol. 1996;64(1):211–5.
22. Verfaillie T, Garg AD, Agostinis P. Targeting ER stress induced apoptosis and inflammation in cancer. Cancer Lett. 2013;332(2):249–64.
23. Taylor CT, McElwain JC. Ancient atmospheres and the evolution of oxygen sensing via the hypoxia-inducible factor in metazoans. Physiology. 2010;25(5):272–9.
24. Hamblin MR, Hasan T. Photodynamic therapy: a new antimicrobial approach to infectious disease? Photochem Photobiol Sci. 2004;3(5):436–50.
25. Simic MG, Jovanovic SV. Antioxidation mechanisms of uric acid. J Am Chem Soc. 1989;111(15):5778–82.
26. Sheehan D, Meade G, Foley VM, Dowd CA. Structure, function and evolution of glutathione transferases: implications for classification of non-mammalian members of an ancient enzyme superfamily. Biochem J. 2001;360(Pt 1):1–16.
27. Lafleur MV, Hoorweg JJ, Joenje H, Westmijze EJ, Retel J. The ambivalent role of glutathione in the protection of DNA against singlet oxygen. Free Radic Res. 1994;21(1):9–17.
28. Miller AC, Henderson BW. The influence of cellular glutathione content on cell survival following photodynamic treatment in vitro. Radiat Res. 1986;107(1):83–94.
29. Wang HP, Qian SY, Schafer FQ, Domann FE, Oberley LW, Buettner GR. Phospholipid hydroperoxide glutathione peroxidase protects against singlet oxygen-induced cell damage of photodynamic therapy. Free Radic Biol Med. 2001;30(8):825–35.
30. Oberdanner CB, Plaetzer K, Kiesslich T, Krammer B. Photodynamic treatment with fractionated light decreases production of reactive oxygen species and cytotoxicity in vitro via regeneration of glutathione. Photochem Photobiol. 2005;81(3):609–13.
31. Genina EA, Bashkatov AN, Sinichkin YP, Yanina IY, Tuchin VV. Optical clearing of biological tissues: prospects of application in medical diagnostics and phototherapy. J Biomed Photon Eng. 2015;1(1):22–58.
32. Huang YY, Vecchio D, Avci P, Yin R, Garcia-Diaz M, Hamblin MR. Melanoma resistance to photodynamic therapy: new insights. Biol Chem. 2013;394(2):239–50.
33. Mroz P, Huang YY, Szokalska A, Zhiyentayev T, Janjua S, Nifli AP, et al. Stable synthetic bacteriochlorins overcome the resistance of melanoma to photodynamic therapy. FASEB J. 2010;24(9):3160–70.
34. Casas A, Di Venosa G, Hasan T, Batlle A. Mechanisms of resistance to photodynamic therapy. Curr Med Chem. 2011;18(16):2486–515.
35. Kishen A, Upadya M, Tegos GP, Hamblin MR. Efflux pump inhibitor potentiates antimicrobial photodynamic inactivation of Enterococcus faecalis biofilm. Photochem Photobiol. 2010;86(6):1343–9.

36. Prates RA, Kato IT, Ribeiro MS, Tegos GP, Hamblin MR. Influence of multidrug efflux systems on methylene blue-mediated photodynamic inactivation of Candida albicans. J Antimicrob Chemother. 2011;66(7):1525–32.
37. Tegos GP, Masago K, Aziz F, Higginbotham A, Stermitz FR, Hamblin MR. Inhibitors of bacterial multidrug efflux pumps potentiate antimicrobial photoinactivation. Antimicrob Agents Chemother. 2008;52(9):3202–9.
38. Feijo Delgado F, Cermak N, Hecht VC, Son S, Li Y, Knudsen SM, et al. Intracellular water exchange for measuring the dry mass, water mass and changes in chemical composition of living cells. PLoS One. 2013;8(7):e67590.
39. Gryson O. Servier medical art France: Servier; 2016. Available from: http://www.servier.com/Powerpoint-image-bank.

Chapter 6
Systemic Effects

Michael Richard Hamblin and Caetano Padial Sabino

Abstract Photodynamic therapy (PDT) is a clinically approved practice for treatment of cancer and infectious diseases. PDT involves systemic or topical administration of a photosensitizer (PS), followed by irradiation of the target area with light of a wavelength matching the absorption band of the PS. In the presence of oxygen, photochemical reactions trigger the production of reactive oxygen species and, consequently, cell death by oxidative stress. Besides causing direct cytotoxicity to tumor cells, PDT induces destruction of the tumor vasculature releasing pro-inflammatory cytokines. Current literature supports that PDT is able to affect both the innate and adaptive responses of the immune system. In addition, PDT-induced adaptive immunity may attack distant untreated tumor cells and lead to development of antitumor memory immunity, which can potentially avoid the cancer relapse. Conversely, pro-inflammatory activity of PDT can also collaborate to resolve local infections since more neutrophils are recruited to the infected region.

M.R. Hamblin (✉)
Wellman Center for Photomedicine, Massachusetts General Hospital,
50 Blossom Street, Bartlett Hall, Room 414, Boston, MA, USA

Department of Dermatology, Harvard Medical School, Boston, MA, USA

Harvard-MIT Division of Health Sciences and Technology, Cambridge, MA, USA
e-mail: hamblin@helix.mgh.harvard.edu

C.P. Sabino
Department of Microbiology, Institute for Biomedical Sciences, University of São Paulo,
Av. Lineu Prestes 1347, Cidade Universitária, Sao Paulo 05508-000, SP, Brazil

Department of Clinical Analysis, School of Pharmaceutical Sciences, University of São Paulo, Sao Paulo, Brazil

Center for Lasers and Applications, Nuclear and Energy Research Institute, National Commission for Nuclear Energy, Sao Paulo, SP, Brazil

Department of Medical Biophysics, Princess Margaret Cancer Institute, University of Toronto, Toronto, Canada
e-mail: caetanosabino@gmail.com

F.P. Sellera et al. (eds.), *Photodynamic Therapy in Veterinary Medicine: From Basics to Clinical Practice*, DOI 10.1007/978-3-319-45007-0_6

6.1 Introduction

When photosensitizer (PS) preparations are injected into the bloodstream, they tend to bind to various serum proteins, and this binding can dramatically affect their pharmacokinetics and biodistribution [1]. Different PSs can have very different pharmacokinetics and this can directly affect the illumination parameters and particularly the pharmacokinetics will determine the drug-light interval [2]. Intravenously injected PSs undergo a gradual transition from being bound to serum proteins, then becoming bound to endothelial cells, and then they are bound to the adventitia of the vessels, then bound either to the extracellular matrix or to the cells within the tumor. Eventually, they will be cleared from the tumor tissue by removal by lymphatic system or by blood vessels, and they will be excreted either by the kidneys or the liver. The effect of PDT on the tumor largely depends at which stage of this continuous process light is delivered.

The antitumor effects of PDT are divided into three main mechanisms. Direct tumor cell death by apoptosis or necrosis can occur, if the PS has been allowed to be taken up by tumor cells. Powerful anti-vascular effects can lead to thrombosis and hemorrhage in tumor blood vessels that subsequently lead to shutdown of tumor blood vessels, which kill the tumor by deprivation of oxygen and nutrients. Finally the release of damage-associated molecular patterns (DAMPs) leads to acute inflammation and the release of cytokines [3]. The DAMPs, cytokines, and stress-response proteins induced in the tumor by PDT work together and lead to an influx of leukocytes that can both contribute to tumor destruction and to stimulate the immune system to recognize and destroy tumor cells even at distant locations.

6.2 PS Pharmacokinetics

When PS is injected into the bloodstream, a certain sequence of events commences that can take very different lengths of time to reach completion for PS with different chemical structures. Firstly, depending on the delivery solvent or vehicle that is used for the PS injection, the PS must come to equilibrium with components of circulating blood. This can involve the PS disaggregating from itself, being released from its delivery vehicle and binding instead to various protein components of serum (see later). In addition the blood has many circulating cells (erythrocytes and leukocytes) that would in principle be available to bind injected PS [4]. Secondly, the circulating PS must bind to the walls of the blood vessels, and it is thought that the nature of the various blood vessels in the tumor and in surrounding normal tissues varies significantly in size, speed of blood flow, and physiological characteristics, such as leakiness and tortuosity in tumors and various normal organs, and governs to a great extent where the PS localizes or accumulates. Thirdly, the PS will penetrate through the wall of the blood vessel at a rate that probably depends on how strong the initial binding of the PS to the intimal surface was. PSs that bind strongly to the blood vessel wall will take a longer time to cross the entire wall of the blood vessel, and those that have an initial weaker binding will pass through more quickly. Fourthly, after extravasation, the PS will diffuse throughout the parenchyma of the organ or tumor to which it has been delivered. If the organ happens to be the liver or other metabolically active organ, the PS may be subject to chemical changes by metabolic enzymes, but this is thought to be rather unlikely for most tetrapyrrole-based PS (e.g., porphyrins, chlorins, and phthalocyanines) in common clinical use. Fifthly, the PS will eventually be eliminated from the tissue. Although this elimination has not been much studied, if the tissue containing the PS is a tumor or other non-metabolic organ, elimination will be probably by lymphatic drainage, and the PS will then return into the general blood circulation via the thoracic duct. Sixthly, the PS will be excreted from the body. For the majority of PS in clinical use, excretion is from the liver into the bile and thence to the intestine where it is lost via fecal elimination. However, enterohepatic recirculation, which is the term used for the reabsorption of the PS from the small intestine back into the bloodstream, has been reported for some PS.

Measurement of pharmacokinetics requires sequential measurements of PS concentration in tissue or other biological substance or fluid in order to construct temporal profiles that can be compared for different PSs. This is frequently done for blood where samples can be readily removed in small amounts at frequent intervals and analysis and then yields values of serum half-lives (first or second order) in minutes, hours, or days. Alternatively, analysis of PS concentrations in urine or fecal samples can give terminal elimination half-lives. All these methods require a method of quantitatively analyzing PS concentrations in samples of biological material. The fact that the majority of PSs used for PDT are also fluorescent has led to the development of assays involving homogenization or dissolving the tissue or biological fluid and measurement of fluorescence in the solution after centrifugation or extraction steps [5]. Calibration curves can be prepared with mixtures of known amounts of PS and weighed amounts of biological material. Care must be taken when these experiments are carried out on small rodents such as mice and rats because these animals accumulate a naturally red fluorescent compound derived from chlorophyll in their diet that ends up in the skin and the digestive tract among other organs [6]. This can be avoided by feeding the animals a chlorophyll-free diet for enough time before the experiment to allow the interfering fluorescent compound to be eliminated [6]. In some cases, the PS is quantified by a traditional analytical chemistry technique such as high-performance liquid chromatography [7]. This method has the advantage of detecting possible metabolism products that have been formed from the originally pure PS after injection. Finally in some cases of experimental animal models, the PS has been presynthesized in the laboratory to include a radioisotope label (for instance, carbon 14 or tritium), or a chelator can be attached that can bind a metal radioisotope, and biological samples can then be quantified using a scintillation counter [7–9]. An alternative approach to carrying out analysis of samples of biological material removed sequentially is to use a continuous optical monitoring method to determine PS concentration in tissue based on fluorescence. In this approach, a noninvasive or minimally invasive approach with fiber-based fluorescence detection systems can be used to measure the variation over time of a fluorescence signal linked to the PS of interest [10]. In some small animal models, the skin can be removed from a subcutaneous tumor so the fiber that collects fluorescence can be in contact with the underlying tumor as well as the adjacent skin [11].

There has been a wide variation reported in pharmacokinetic parameters for various PSs that are in clinical or preclinical use. Bellnier and Dougherty [12] studied pharmacokinetics of Photofrin® in patients scheduled to undergo PDT for the treatment of carcinoma of the lung or the skin. They found a triexponential 3-compartment pharmacokinetic model with alpha, beta, and gamma half-lives of approximately 16 h, 7.5 days, and 155.5 days. Detectable Photofrin® fluorescence was shown to persist in the serum for longer than 1 year. In clinical practice, PDT light is usually delivered to patients 48 h (drug-light interval) after Photofrin® injection. In a Japanese study [13], skin photosensitivity was found to persist for as long as a month or more after Photofrin® injection and interestingly was significantly more pronounced for female patients.

The pharmacokinetics of 2-[1-hexyloxyethyl]-2-devinyl pyropheophorbide-a (HPPH) has been studied in cancer patients [14]. A two-compartment model yielded alpha and beta half-lives of 7.77 h and 596 h. Radiolabeled Foscan® pharmacokinetics were studied in tumor-bearing rats yielding a triexponential model with alpha, beta, and gamma half-lives of 0.46 h, 6.91 h, and 82.5 h, respectively [15]. Pharmacokinetics of the silicon phthalocyanine Pc4 were studied in non-tumor-bearing mice, and a two-compartment fit was obtained with alpha and beta half-lives of approximately 10 min and 20 h with some variation depending on the injected dose and solvent [16]. The palladium bacteriopheophorbide PS known as TOOKAD® has very rapid pharmacokinetics with alpha and beta half-lives of approximately 2 min and 1.3 h, and in this case, graphite furnace atomic absorption spectroscopy was used to quantify the palladium atom coordinated in the center of the tetrapyrrole [17].

6.3 PS Biodistribution

In experimental animals, it has sometimes been possible to establish the overall pattern of distribution in the different organs after intravenous (IV) injection of the PS. A study using radioisotope labeled Photofrin® in mice with subcutaneous fibrosarcoma tumors that were sacrificed 24 h later found the highest concentrations in the liver, adrenal glands, and urinary bladder [7]. Next were the pancreas, kidney, and spleen, and these organs were higher than the stomach, bone, lung, and heart. Muscle concentration was low and the brain was lowest. Only the skeletal muscle, brain, and skin located distant to the tumor had peak concentrations lower than tumor tissue; the skin overlying the tumors showed concentrations not significantly different from tumor. The higher molecular weight components of Photofrin® were partially retained in the liver and spleen at 75 days after injection.

In most cases where organ biodistribution data have been reported for small animals such as mice and rats, the highest concentration of PS has been found in the liver [19, 20]. Chan et al. studied a set of sulfonated phthalocyanines and found that the accumulation in the liver was inversely proportional to the degree of sulfonation and hence to the overall lipophilicity of the PS molecules [21]. The liver is known to have a highly permeable blood supply with fenestrated endothelium that contains pores that allow molecules to pass easily out of the blood vessels. This porous vasculature is required by the role of the liver in detoxifying the body of extraneous molecules (particularly organic molecules such as PS). These molecules are then excreted into the bile via the gall bladder in a similar fashion to the production of bile acids from cholesterol. The bile is routed into the duodenum where it is possible for unchanged PS to be absorbed into the circulation a second time from the small intestine (a process known as enterohepatic recycling) as well as being excreted via the feces. It is known that soon after injection, large amounts of PS can accumulate in the lungs. Lung accumulation has been reported as the mode of dose-limiting dark toxicity when the PS injected doses were increased eventually leading to lung hemorrhage and acute interstitial pneumonia after injection of a high dose of 80 mg/kg of Pc4 into mice [17]. The explanation is probably that large injected doses of PS can aggregate in the blood vessels and the particles so formed can easily collect in the fine capillary network of the lungs.

The amount of PS that accumulates in the spleen seems to vary widely with the structure of the PS (overall charge and hydrophobicity). Woodburn et al. studied a range of porphyrins with varying octanol/water partition coefficients and found that the accumulation in the spleen varied the most compared to other organs [18]. Egorin et al. compared organ distribution of Pc4 administered by IV injection of the same dose (10 mg/kg) dissolved in several solvents and found that the accumulation in the spleen was the most variable [16].

The kidney and bladder frequently accumulate high amounts of PS, despite the clearance route being almost exclusively via the liver and bile [19] rather than the kidneys and urine. The digestive organs (stomach, large and small intestines) seem to take up middling amounts of PS (i.e., less than liver but more than muscle) [9]. The skin has not been found to accumulate high amounts of PS in experimental animals, which might be considered somewhat surprising in view of the fact that skin photosensitivity due to PS accumulation in patients' skin is the main cause of PDT-related side effects. The lowest concentrations of PS are generally found in organs such as the heart, skeletal muscle, bone, eye, and brain. These are organs that are known to have a relatively impermeable blood supply. The blood-brain barrier serves to exclude PS from the parenchyma of the brain, while the fact that the blood-brain barrier is frequently breached by the growth of brain tumors is probably responsible for the very large (over 100) tumor-to-normal brain ratios reported for some PS [1].

Since the original observation of the tumor-localizing ability of hematoporphyrin derivative (HPD), many workers have investigated the underlying mechanism of this effect whereby PSs preferentially localize in tumors and in other specific organs and anatomical locations [20–22]. The precise definition of "tumor-to-normal tissue ratio" has also been subject to debate. Some investigators working with subcutaneous tumors in experimental animals use the ratio between the tumor and peritumoral muscle or skin, while others use the ratio between the tumor and the distant muscle and skin. Furthermore, there has been much effort made to determine which molecular features in the chemical structures of the PS are optimal for maximizing the selectivity for the tumor over normal tissue and organs. This has proved quite complicated because the pharmacokinetics of different PS can vary dramatically. For instance, one PS can have its best tumor-to-normal tissue ratio at a relatively early time point after administration such as 3 h, while for another PS, this time point can be as long as 7 days after injection. One of the notable properties of these tetrapyrrole molecules that are commonly employed as PS, which may be relevant to their tumor localizing ability, is their tendency to bind strongly to serum proteins and also to bind to each other (aggregation). This high tendency to bind means that most PSs when injected into the bloodstream behave as macromolecules rather than as small molecules, either because they are more or less firmly bound to large protein molecules or because they have formed intermolecular aggregates of similar size. Many works have reported on the distribution of PS between various classes of serum proteins (either human or animal) in vitro [23–25]. These serum proteins are usually divided into four classes: (a) albumin and other heavy proteins, (b) high-density lipoprotein (HDL), (c) low-density lipoprotein (LDL), and (d) very low-density lipoprotein (VLDL). However, even this study has been complicated by the fact that the most lipophilic PSs are insoluble in aqueous media and need to be delivered in a solvent mixture, which may alter the serum protein distribution of the PS. It has been argued that PSs that preferentially bind to LDL are better tumor localizers (see below) [26], but this might not always be the case [27].

It is useful to make a distinction between selective accumulation and selective retention. The tumor localizing ability of the PS with the faster pharmacokinetics is probably due to selective accumulation in the tumor, while the localization of PS with slower-acting pharmacokinetics is more likely due to selective retention. In the selective accumulation model, it is thought that the increased vascular permeability to macromolecules typical of tumor neovasculature is chiefly responsible for the preferential extravasation of the PS. These quick-acting PSs frequently bind to albumin, which is of ideal size, and Stokes radius to pass through the "pores" in the endothelium of the tumor microvessels [28]. The selective retention model where PS can be retained in tumors, while they are eliminated from surrounding normal tissue and organs, has been the subject of much speculation. As mentioned above, a popular theory maintains that the binding of the PS to LDL is of major importance [26]. According to this theory, it is proposed that cancer cells overexpress the LDL (apoB/E) receptor. Upregulation of the expression of LDL receptors is one strategy that rapidly growing malignant cells have developed to obtain sufficient cholesterol that the tumor cells need for the biosynthesis of lipids required due to the rapid turnover of their cellular membranes. There is experimental evidence both for and against this LDL receptor theory [27, 29]. Other theories have been proposed to account for the selective retention of PS in tumor tissue. One is that tumors are characterized by poorly developed lymphatic drainage and that macromolecules, which extravasate from the hyperpermeable tumor neovasculature, are retained in the extravascular space. This theory has been termed the well-known "enhanced permeability and retention" (EPR) effect. Another theory is attributed to macrophages, which are phagocytic cells that infiltrate solid tumors to varying extents. These tumor-associated macrophages (TAMs) have been shown to accumulate up to 13 times the amount of some PS compared to the cancer cells themselves. The explanation for this preferential uptake by TAMs has been proposed to be due to either the phagocytosis of aggregates of PS or to the preferential uptake by TAMs of lipoproteins, which have been subtly altered by the binding of porphyrins. Another theory proposes that the low pH commonly found in tumors (due to the Warburg effect in which the metabolism changes from oxidative phosphorylation to glycolysis and produces large amounts of lactic acid) has the effect of trapping some of the anionic PS, which are normally ionized at normal physiological pH. These PSs then become neutrally charged (protonated) at low pH and hence more lipophilic, when they encounter the acidic pH in the tumor environment [30]. Yet another proposed mechanism is the selective accumulation of PS in tumors related to the propensity of tumors to accumulate lipid droplets which may contain lipophilic molecules [31]. A final theory concerns the tendency of some tumors to have very high populations of tumor-associated macrophages that may engulf particles of aggregated PS [32], or else take up PSs that have bound to LDL, and then altered its behavior to be recognized by scavenger receptors [20].

6.4 Direct Destruction of Tumors by PDT

Many of the studies relating to this topic have been concerned with attempts to distinguish between direct tumor cell killing in vivo and indirect tumor cell killing caused by the vascular shutdown effects of PDT (see later).

Some studies have investigated an experimental model in which tumors in mice that had been treated with a potentially curative dose of PDT were removed from the mice and then dissociated into a suspension of individual tumor cells. These tumor cells were still able to form colonies in a clonogenic assay if the tumors were removed from the host immediately after PDT [33, 34]. However, when these tumors were left in the host animal for increasing lengths of time, a progressive loss in clonogenicity of the dissociated tumor cells was observed. These times corresponded to the emergence of PDT-induced tumor hypoxia as determined radiologically. These studies suggested a central role for vascular damage in governing the tumor response to PDT in murine models.

6.5 Anti-vascular Effect of PDT

Photodynamic alteration of tissue microcirculation was first reported in 1963 [35]. A study by W. M. Star et al. utilized a window chamber in rats bearing an implanted mammary tumor, to make direct observations of the microcirculation of the tumor, and the adjacent normal tissue before, during, and at various times after PDT mediated by HPD [36]. An initial slowing down of blood flow and vasoconstriction of the tumor vessels was followed by heterogeneous responses including eventual complete cessation of blood flow, hemorrhage, and, in some larger vessels, the formation of platelet aggregates. Other workers have histologically examined tissues removed from animal models that had been treated with PDT, such as subcutaneous urothelial tumors and normal rat jejunum range, and reported a range of vascular responses including disruption of blood flow, breakdown of the blood-brain barrier in the normal brain of mice, and endothelial cell damage in subcutaneous tumors and normal tissue [37, 38]. Many reports directly implicate the endothelium as a primary target for PDT in vivo; this stimulated research into the relative sensitivity of endothelial cells to PDT and the responses of endothelial cells that could initiate the various phenomena at the vessel level. Gomer et al. showed that bovine endothelial cells were significantly more sensitive to Photofrin®-mediated PDT than smooth muscle cells or fibroblasts from the same species [39]. This increased sensitivity to PDT, measured by clonogenic assay, was not a result of increased Photofrin® accumulation. Sensitivity to HPD-mediated PDT of bovine aorta endothelial cells and human colon adenocarcinoma cells was investigated by West et al. [40]. Exponentially growing endothelial cells were significantly more sensitive than tumor cells with a similar rate of proliferation, and the difference in sensitivity was accompanied by greater PS accumulation in the endothelial cells.

Intravital microscopy of the rat cremaster muscle showed that Photofrin®-PDT caused vessel constriction with an increase in vessel permeability and leukocyte adhesion was observed [41]. The cyclooxygenase inhibitor indomethacin was able to inhibit the vessel constriction and reduce the increase in vascular permeability, implicating in the involvement of the metabolites of cyclooxygenase called eicosanoids (e.g., thromboxane or prostacyclin). An increase in vascular permeability, thrombus formation, and blood flow reduction were observed in animals treated with verteporfin-PDT in a dose- and time-dependent manner [42–44]. Blood vessels with lower flow rates were more sensitive to verteporfin-PDT-induced vascular shutdown than vessels with a higher flow rate. In agreement with light and electron microscopic studies, intravital fluorescence microscopic imaging demonstrated in live animals that verteporfin-PDT decreased endothelial cell viability and caused platelet aggregation [45]. Rapid shutdown of tumor blood flow by formation of blood clots was also found after PDT with TOOKAD® and MV6401 (indium chloride methyl pyropheophorbide), which was also confirmed by histological assessment [46, 47]. Many optical imaging studies have led to the conclusion that PDT is essentially a vascular disrupting therapy.

6.6 Immune Stimulation Effect of PDT: Innate Immunity

PDT frequently provokes a strong acute inflammatory reaction in the treated site. This can often be observed as a fairly rapid induction of localized edema [48]. Such immune reaction is a direct consequence of PDT-induced oxidative stress. Thus,

PDT can be ranked among cancer therapies (including cryotherapy, hyperthermia, and focused ultrasound ablation) that all produce a chemical or physical destruction of tumor tissue that is perceived by the host as localized acute trauma. This trauma prompts the host to launch protective actions designed by evolution to react against threats to tissue integrity and to try to restore homeostasis at the affected site [49]. The acute inflammatory response is the principal process engaged in this context. Its main task is containing the disruption of homeostasis, ensuring removal of damaged cells, and protection against possible invasion of pathogenic microorganisms that may gain entry to the body as a result of a traumatic breach in external barriers. The inflammatory response also contributes to the local healing process with restoration of normal tissue function. The inflammation elicited by PDT is orchestrated by the innate immune system [49]. The principal cells involved in this process consist of neutrophils, monocyte/macrophages, dendritic cells, and mast cells. The recognition systems of the innate immune system are based upon a wide array of pattern recognition receptors (PRRs), which are responsible for detecting the presence of tissue damage revealed to its sensors as the appearance of "altered self" [49]. As presented in Fig. 6.1, PDT appears particularly effective in generating rapidly an abundance of alarm/danger signals, also called DAMPs, at the treated site that can be detected by the innate immune cells [49]. Many DAMPs are nuclear or cytosolic proteins or even small molecules that are normally safely contained inside healthy cells. When released outside from damaged or necrotic cells, or exposed on the surface of the cell following damage, they can encounter an oxidizing environment, which results in their changes and in their conformation or chemical structure [50]. DNA can be released outside the nucleus from necrotic cells and becomes a DAMP [51]. Other DAMPs include heat shock proteins, high-mobility group box protein 1 (HMGB1), hyaluronan fragments, extracellular ATP, uric acid, heparin sulfate, and S100 molecules (members of a family of calcium-modulated proteins, Fig. 6.1) [50].

Dendritic cells (DCs) are professional antigen-presenting cells that have been described as the "orchestrators" of the immune response [52]. DCs express many receptors that specifically recognize DAMPs. Since PDT of tumor cells causes both cell death and cell stress [48, 53, 54], it is hypothesized that the activation of DCs by PDT-treated cells is the result of recognition of DAMPs released/secreted/exposed by PDT from dying cells (see Fig. 6.1) [55–57]. HSP70 is a well-characterized DAMP that interacts with the danger signal receptors, TLRs (Toll-like receptors) 2 and 4 [58], and is induced by PDT [59]. The level of expression of HSP70 in PDT-treated tumor cells appears to correlate with an ability to stimulate DC maturation [60] and initiation of inflammation [56, 61].

The onset of PDT-induced inflammation is marked by dramatic changes in the tumor vasculature that becomes permeable to serum proteins and expresses adhesion molecules that circulating inflammatory cells can recognize and to which they can adhere [49]. This process is responsible for the well-known "rolling" behavior of neutrophils and other leukocytes. The inflammatory process is predominantly initiated by signals originating from photooxidative damage produced in perivascular regions of the tumor with chemotactic gradients reaching out from the damaged endothelium. The inflammatory cells initially consisting of neutrophils, which are

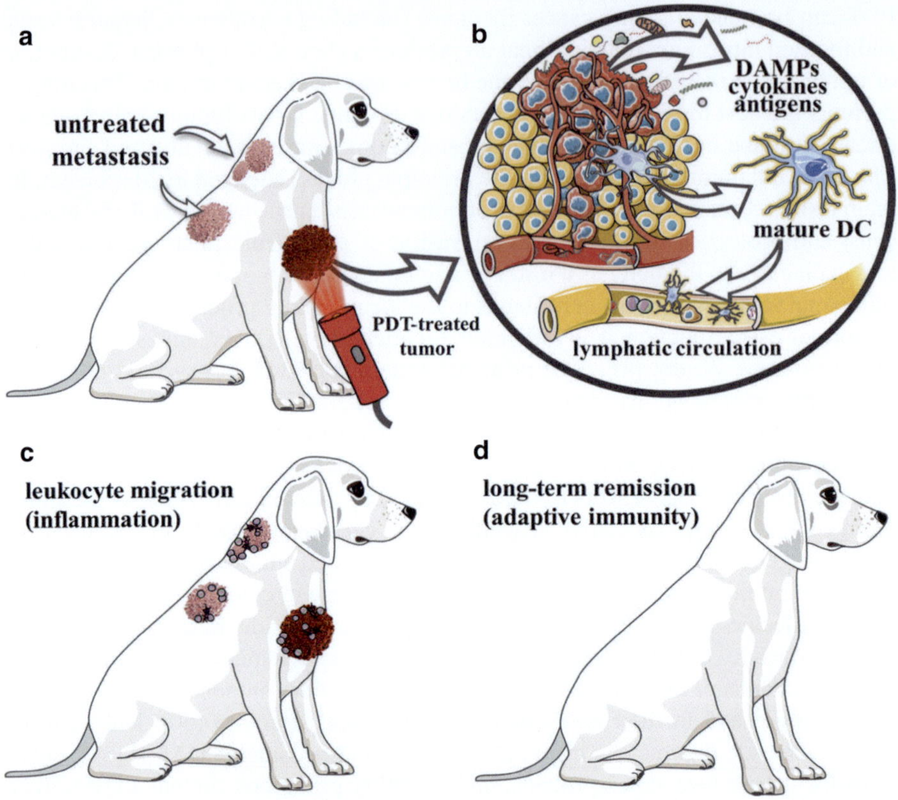

Fig. 6.1 Schematic illustration of antitumor immune response triggered by PDT. Immunogenic tumor cell death caused by PDT (**a**) induces abundant release of danger-associated molecular patterns (DAMPs) at the treated site that can be detected by antigen-presenting cells, such as dendritic cells (DCs). Following activation, DCs migrate through lymphatic vessels to the closest lymph node to prime the adaptive immune response via clonal expansion of lymphocyte capable to recognize tumor-specific antigens (**b**). After adaptive immunity is formed, cytotoxic T cells along with other leukocytes roam through the organism to combat distant tumor metastasis (**c**) and promote a long-term remission (**d**) [98]

then followed by mast cells and monocytes/macrophages, rapidly and massively invade tumors after PDT [48, 62]. The primary task of these infiltrating cells is to neutralize the source of DAMPs by removing debris containing injured and dead cells and oxidatively damaged tissue components. Damage and dysfunction of PDT-treated tumor vasculature frequently results in vascular occlusion that serves to "seal off" the damaged tumor tissue until it is removed by invading phagocytic cells [49]. Depletion of neutrophils or inhibition of their activity after PDT was shown to diminish the therapeutic effect [63–66]. Among the cytokines involved in the regulation of the inflammatory process, the most important ones for the PDT response are IL-1β and IL-6 [67, 68]. Blocking the function of various adhesion molecules was also shown to inhibit the PDT response [67, 68]. On the other hand, blocking anti-inflammatory cytokines such as IL-10 and TGF-β can significantly improve the cure rates after PDT [49].

6.7 Immune Stimulation Effect of PDT: Adaptive Immunity

A wide range of preclinical animal studies and even some clinical studies have demonstrated that PDT can influence the adaptive immune response in disparate ways. Some PDT regimens result in potentiation of adaptive immunity, while others lead to immunosuppression. The precise reasons why PDT can lead to potentiation vs. suppression are unclear; however, it appears as though the effect of PDT on the immune system is dependent upon the treatment regimen, as well as the area treated and the precise PS type [65, 69]. PDT-induced immunosuppression is largely confined to cutaneous and transdermal PDT regimens involving large surface areas [69, 70].

Broadly speaking PDT efficacy appears to be dependent upon the induction of adaptive antitumor immunity (see Fig. 6.1). Long-term tumor response and cures are diminished or absent in immunocompromised mice such as nude mice or SCID mice [63, 71]. Reconstitution of these animals by injecting bone marrow or T cells isolated from immunocompetent (normal) mice can result in increased PDT efficacy. Canti et al. [72] were the first to show PDT-induced immune stimulation, demonstrating that cells isolated from tumor-draining lymph nodes of PDT-treated mice were able to confer tumor resistance when injected into naïve mice. Subsequent studies demonstrated that PDT directed against murine tumors resulted in the generation of immune memory such as rejection of a tumor rechallenge in mice that had obtained a long-term cure after PDT [73].

For the adaptive immune response produced by PDT to be highly effective, it appears that the tumor cell line employed should express what has been called as "tumor rejection antigen." This has been demonstrated by testing PDT of syngeneic mouse tumors that have been engineered to express a model tumor antigen in immunocompetent mice. One of these model antigens was the bacterial protein beta-galactosidase (b-gal) [74]. Although the b-gal tumors grew as rapidly as their wild-type counterparts, there were 100 % long-term cures after liposomal benzoporphyrin derivative (BPD)-mediated PDT with the b-gal tumors but no long-term cures with the wild-type tumors. The strength of the antitumor immune response was demonstrated by the fact that if mice had two b-gal tumors (one in each leg) and only one was treated with PDT, the contralateral untreated tumor underwent spontaneous regression in 70 % of the mice. A similar result was seen in mice bearing the mastocytoma P815 that is well known to express the naturally occurring tumor rejection antigen called P1A [75]. The CD8 cytotoxic T cells can recognize specific peptide epitopes derived from the amino acid sequence of the tumor antigen.

Major histocompatibility complex class I molecules (MHC-I) is critical for allowing the $CD8^+$ T cells to recognize the tumor antigen, and tumors that lack MHC-I are resistant to cell-mediated antitumor immune reactions [76]. Since MHC-1 expression and expression of tumor-associated antigens (such as P1A) are both susceptible to downregulation by epigenetic silencing, an epigenetic reversal agent (the demethylating compound, 5-aza-2′-deoxycytidine) could potentiate the induction of antitumor immunity and increase cures in several mouse tumor models [77].

The mechanism whereby PDT enhances antitumor immunity has been examined for the past several decades. PDT activates both humoral and cell-mediated antitumor immunity, although the importance of the humoral response (antibodies) is at present unclear. PDT efficacy in mice and humans is reduced in the absence of $CD8^+$ T cell activation and/or tumor infiltration [63, 78, 79]. Therefore, most mechanistic studies have focused on the means by which PDT potentiates $CD8^+$ T cell activation. It is clear that induction of antitumor immunity following PDT is dependent upon induction of inflammation [80]. PDT-induced acute local and systemic inflammation is postulated to culminate in the maturation and activation of DCs. Mature DCs are critical for the activation of tumor-specific $CD8^+$ T cells and induction of antitumor immunity [81]. DCs are activated in response to PDT [68] and migrate to tumor-draining lymph nodes (see Fig. 6.1) where they present the antigens they have taken up from damaged tumor cells to naïve T cells in the so-called immune synapse [82]. The T cells in the lymph node become activated and proliferate dramatically, and a "flood" of tumor-specific CTLs is released into the bloodstream where they can search for other tumor cells to attack and destroy (see Fig. 6.1) [68, 83]. Generation of $CD8^+$ effector and memory T cells is frequently, but not always, dependent upon the presence and activation of $CD4^+$ T cells [84]. PDT-induced antitumor immunity may [63] or may not depend on $CD4^+$ T cells [79] and may be augmented by natural killer (NK) cells [79].

PDT-mediated enhancement of antitumor immunity is believed to be due, at least in part, to the stimulation of DCs by dead and dying tumor cells, suggesting that in vitro PDT-treated tumor cells may act as effective antitumor vaccines [85]. This hypothesis has been proven by several studies using a wide variety of different PSs and tumor models in both preventative and therapeutic settings [66, 85–87].

Mechanistic studies showed that incubation of immature DCs with PDT-treated tumor cells leads to enhanced DC maturation, activation, and increased ability to stimulate T cells [85, 88]. Furthermore, opsonization of PDT-treated tumor cells by complement proteins increases the efficacy of PDT-generated vaccines [86]. The implications of PDT-induced antitumor immunity and efficacious PDT-generated vaccines are significant and provide an exciting possibility for using PDT in the treatment of metastatic disease and as an adjuvant in combination with other cancer modalities. Several preclinical studies demonstrated that PDT is able to control the growth of tumors present outside the treatment field [79, 89] although others have failed to demonstrate control of distant disease following PDT [90].

The efficacy of clinical PDT also appears to depend on induction of antitumor immunity. Patients with vulval intraepithelial neoplasia (VIN) who did not respond to aminolevulinic acid (ALA)-PDT were more likely to have tumors that lacked major histocompatibility complex class I molecules (MHC-I) than patients who responded to ALA-PDT [78]. VIN patients who responded to PDT had increased $CD8^+$ T cell infiltration into the treated tumors as compared to nonresponders. In other clinical studies, patients who were treated with PDT for actinic keratoses and Bowen's disease were divided into immunocompetent and immunosuppressed subgroups. Both subgroups had similar initial response rates to PDT, but immunosuppressed patients had less long-term responses and were more likely to suffer the appearance of new lesions [91]. PDT of multifocal angiosarcoma of the head and neck resulted in increased immune cell infiltration into distant untreated tumors that was accompanied by tumor regression [92, 93]. PDT was also shown to be an effective surgical adjuvant in non-small cell lung cancer patients with pleural spread [94]. Recent reports have shown that clinical antitumor PDT also increases antitumor immunity. PDT of basal cell carcinoma (BCC) increased immune cell reactivity against a BCC-associated antigen [95].

6.8 Antimicrobial Photodynamic Therapy (APDT) for Infectious Disease

In recent years, the application of APDT for the treatment of localized infections (also called antimicrobial photodynamic inactivation, or aPDI) has steadily increased in popularity. This application can superficially appear much simpler than the more well-recognized application of PDT for cancer treatment. The complexity of dealing with the vascular effects of PDT is absent, because the PS is almost always topically applied into/onto the infected area of tissue. Moreover, current literature is still very limited about activation of the immune system triggered by PDT for infectious diseases (see later).

So it is generally assumed that a PS with cationic charges, which is more or less selective for microbial cells, is injected or otherwise topically applied into the infected area where it binds to the microbial cells after a short drug-light interval, in order to reduce the uptake of the PS by the surrounding host cells, and increases selectivity. However, what is not often realized is the big difference between tumors and infections. In tumors all the surrounding tissue is considered to be undesirable, diseased, and in need of removal (e.g., surgical margins). This is completely different to the case of an infection. An infection is defined as greater than 100,000 colony-forming units (CFU) of microbial cells per gram of tissue. The mass of 10^5 CFU of bacterial cells is approximately 1 μg. So the ratio between the mass of the bacterial tissue and the mass of the host tissue is in the region of 1–1,000,000. So in sharp contrast to the case, where in tumors the goal is to destroy 100 % of the surrounding tissue, in infections the goal is to destroy only 0.0001 % of the surrounding tissue (just the bacterial cells). For this to be anyway near feasible, the selectivity of the PS for bacteria has to be exquisite so that 999,999 PS molecules out of every million bind to their target bacteria and not to the host cells. As can be imagined, this is rather unlikely. However, the situation is not as dire as the foregoing has made it out to be. The reason for this is that a short drug-light interval is used for APDT, and even if the selectivity for bacteria over host cells is somewhat less than 1 million to 1, then the polycationic nature of the PS means that it is fairly slow to be taken up by host mammalian cells. So if light is delivered after say 30 s from topical delivery of the PS, it can be reasonably safe to assume that not many of the surrounding host cells would be killed.

There have not been many reports of damage to host tissue being caused by APDT in animal models. One study used a model of a subcutaneous abscess caused by Gram-positive bioluminescent *Staphylococcus aureus* injected into the mouse thigh muscle [96]. A small volume of PS solution was injected into the infected area, and light was shone as a spot onto the surface. Two different PS were compared, the polycationic conjugate, poly-L-lysine chlorin(e6) (pL-ce6), and the anionic PS, free chlorin(e6). Both PSs were equally effective in destroying the bacteria as judged by noninvasive bioluminescence imaging. However, the nonbacterial-targeted free chlorin(e6) caused more damage to the surrounding mouse thigh muscle as measured by a leg function test. The bacterial-targeted pL-ce6 was able to kill the bacteria without damaging the thigh muscle.

There has only been one report of activation of the immune system after APDT, when used against an infection in an animal model [97]. This involved injection of bioluminescent *S. aureus* into the mouse knee to produce a form of septic arthritis. When methylene blue was injected into the knee and red light ($\lambda = 660$ nm) delivered to the surface, the bacteria were killed in a similar manner to the experiment described above. However, when the MB-mediated PDT was carried out before the bacteria were injected into the knee, rather than afterward, an interesting observation was made. APDT carried out 24 h (but not 2 h) before infection protected the mice from developing a severe knee infection when the bacteria were subsequently injected, as shown by the fact that the infection resolved earlier [97]. This effect was dependent on neutrophils in the mouse knee as demonstrated by the use of several blocking antibodies and inhibitors. Although there was no immediate influx of neutrophils into the mouse knee immediately after APDT, the neutrophils did appear to be "primed" to react quicker to attack the bacteria, once these were injected into the knee joint. Therefore, PDT of the mouse knee could be envisaged as a "nonspecific vaccination procedure," which could in principle be used in cases of orthopedic surgery, which might be considered high risk for infection.

Acknowledgments MR Hamblin was supported by the US NIH Grant R01AI050875.

References

1. Boyle RW, Dolphin D. Structure and biodistribution relationships of photodynamic sensitizers. Photochem Photobiol. 1996;64(3):469–85.
2. Agostinis P, Berg K, Cengel KA, Foster TH, Girotti AW, Gollnick SO, et al. Photodynamic therapy of cancer: an update. CA Cancer J Clin. 2011;61(4):250–81.
3. Garg AD, Krysko DV, Vandenabeele P, Agostinis P. DAMPs and PDT-mediated photo-oxidative stress: exploring the unknown. Photochem Photobiol Sci. 2011;10(5):670–80.
4. Maugain E, Sasnouski S, Zorin V, Merlin JL, Guillemin F, Bezdetnaya L. Foscan-based photodynamic treatment in vivo: correlation between efficacy and Foscan accumulation in tumor, plasma and leukocytes. Oncol Rep. 2004;12(3):639–45.

5. Bellnier DA, Greco WR, Parsons JC, Oseroff AR, Kuebler A, Dougherty TJ. An assay for the quantitation of Photofrin in tissues and fluids. Photochem Photobiol. 1997;66(2):237–44.
6. Gudgin Dickson EF, Holmes H, Jori G, Kennedy JC, Nadeau P, Pottier RH, et al. On the source of the oscillations observed during in vivo zinc phthalocyanine fluorescence pharmacokinetic measurements in mice. Photochem Photobiol. 1995;61(5):506–9.
7. Bellnier DA, Ho YK, Pandey RK, Missert JR, Dougherty TJ. Distribution and elimination of Photofrin II in mice. Photochem Photobiol. 1989;50(2):221–8.
8. Little FM, Gomer CJ, Hyman S, Apuzzo ML. Observations in studies of quantitative kinetics of tritium labelled hematoporphyrin derivatives (HpDI and HpDII) in the normal and neoplastic rat brain model. J Neurooncol. 1984;2(4):361–70.
9. Schuitmaker JJ, Feitsma RI, Journee-De Korver JG, Dubbelman TM, Pauwels EK. Tissue distribution of bacteriochlorin a labelled with 99mTc-pertechnetate in hamster Greene melanoma. Int J Radiat Biol. 1993;64(4):451–8.
10. Frisoli JK, Tudor EG, Flotte TJ, Hasan T, Deutsch TF, Schomacker KT. Pharmacokinetics of a fluorescent drug using laser-induced fluorescence. Cancer Res. 1993;53(24):5954–61.
11. Sheng C, Pogue BW, Wang E, Hutchins JE, Hoopes PJ. Assessment of photosensitizer dosimetry and tissue damage assay for photodynamic therapy in advanced-stage tumors. Photochem Photobiol. 2004;79(6):520–5.
12. Bellnier DA, Dougherty TJ. A preliminary pharmacokinetic study of intravenous Photofrin in patients. J Clin Laser Med Surg. 1996;14(5):311–4.
13. Moriwaki SI, Misawa J, Yoshinari Y, Yamada I, Takigawa M, Tokura Y. Analysis of photosensitivity in Japanese cancer-bearing patients receiving photodynamic therapy with porfimer sodium (Photofrin). Photodermatol Photoimmunol Photomed. 2001;17(5):241–3.
14. Bellnier DA, Greco WR, Loewen GM, Nava H, Oseroff AR, Pandey RK, et al. Population pharmacokinetics of the photodynamic therapy agent 2-[1-hexyloxyethyl]-2-devinyl pyropheophorbide-a in cancer patients. Cancer Res. 2003;63(8):1806–13.
15. Jones HJ, Vernon DI, Brown SB. Photodynamic therapy effect of m-THPC (Foscan) in vivo: correlation with pharmacokinetics. Br J Cancer. 2003;89(2):398–404.
16. Egorin MJ, Zuhowski EG, Sentz DL, Dobson JM, Callery PS, Eiseman JL. Plasma pharmacokinetics and tissue distribution in CD2F1 mice of Pc4 (NSC 676418), a silicone phthalocyanine photodynamic sensitizing agent. Cancer Chemother Pharmacol. 1999;44(4):283–94.
17. Brun PH, DeGroot JL, Dickson EF, Farahani M, Pottier RH. Determination of the in vivo pharmacokinetics of palladium-bacteriopheophorbide (WST09) in EMT6 tumour-bearing Balb/c mice using graphite furnace atomic absorption spectroscopy. Photochem Photobiol Sci. 2004;3(11–12):1006–10.
18. Woodburn KW, Stylli S, Hill JS, Kaye AH, Reiss JA, Phillips DR. Evaluation of tumour and tissue distribution of porphyrins for use in photodynamic therapy. Br J Cancer. 1992;65(3): 321–8.
19. Richter AM, Cerruti-Sola S, Sternberg ED, Dolphin D, Levy JG. Biodistribution of tritiated benzoporphyrin derivative (3H-BPD-MA), a new potent photosensitizer, in normal and tumor-bearing mice. J Photochem Photobiol B. 1990;5(2):231–44.
20. Hamblin MR, Newman EL. On the mechanism of the tumour-localising effect in photodynamic therapy. J Photochem Photobiol B. 1994;23(1):3–8.
21. Jori G. In vivo transport and pharmacokinetic behavior of tumour photosensitizers. Ciba Found Symp. 1989;146:78–86.
22. Larroque C, Pelegrin A, Van Lier JE. Serum albumin as a vehicle for zinc phthalocyanine: photodynamic activities in solid tumour models. Br J Cancer. 1996;74(12):1886–90.
23. Kessel D, Poretz RD. Sites of photodamage induced by photodynamic therapy with a chlorin e6 triacetoxymethyl ester (CAME). Photochem Photobiol. 2000;71(1):94–6.
24. Kongshaug M, Moan J, Brown SB. The distribution of porphyrins with different tumour localising ability among human plasma proteins. Br J Cancer. 1989;59(2):184–8.
25. Maziere JC, Santus R, Morliere P, Reyftmann JP, Candide C, Mora L, et al. Cellular uptake and photosensitizing properties of anticancer porphyrins in cell membranes and low and high density lipoproteins. J Photochem Photobiol B. 1990;6(1–2):61–8.

26. Jori G, Reddi E. The role of lipoproteins in the delivery of tumour-targeting photosensitizers. Int J Biochem. 1993;25(10):1369–75.
27. Korbelik M. Low density lipoprotein receptor pathway in the delivery of Photofrin: how much is it relevant for selective accumulation of the photosensitizer in tumors? J Photochem Photobiol B. 1992;12(1):107–9.
28. Yuan F, Leunig M, Berk DA, Jain RK. Microvascular permeability of albumin, vascular surface area, and vascular volume measured in human adenocarcinoma LS174T using dorsal chamber in SCID mice. Microvasc Res. 1993;45(3):269–89.
29. Allison BA, Pritchard PH, Levy JG. Evidence for low-density lipoprotein receptor-mediated uptake of benzoporphyrin derivative. Br J Cancer. 1994;69(5):833–9.
30. Pottier R, Kennedy JC. The possible role of ionic species in selective biodistribution of photochemotherapeutic agents toward neoplastic tissue. J Photochem Photobiol B. 1990;8(1):1–16.
31. Freitas I. Lipid accumulation: the common feature to photosensitizer-retaining normal and malignant tissues [news]. J Photochem Photobiol B. 1990;7(2–4):359–61.
32. Korbelik M, Krosl G. Photofrin accumulation in malignant and host cell populations of a murine fibrosarcoma. Photochem Photobiol. 1995;62(1):162–8.
33. Henderson BW, Waldow SM, Mang TS, Potter WR, Malone PB, Dougherty TJ. Tumor destruction and kinetics of tumor cell death in two experimental mouse tumors following photodynamic therapy. Cancer Res. 1985;45(2):572–6.
34. Henderson BW, Fingar VH. Oxygen limitation of direct tumor cell kill during photodynamic treatment of a murine tumor model. Photochem Photobiol. 1989;49(3):299–304.
35. Castellani A, Pace GP, Concioli M. Photodynamic effect of haematoporphyrin on blood microcirculation. J Pathol Bacteriol. 1963;86:99–102.
36. Star WM, Marijnissen HP, van den Berg-Blok AE, Versteeg JA, Franken KA, Reinhold HS. Destruction of rat mammary tumor and normal tissue microcirculation by hematoporphyrin derivative photoradiation observed in vivo in sandwich observation chambers. Cancer Res. 1986;46(5):2532–40.
37. Bhuvaneswari R, Gan YY, Soo KC, Olivo M. The effect of photodynamic therapy on tumor angiogenesis. Cell Mol Life Sci. 2009;66(14):2275–83.
38. Tseng MT, Reed MW, Ackermann DM, Schuschke DA, Wieman TJ, Miller FN. Photodynamic therapy induced ultrastructural alterations in microvasculature of the rat cremaster muscle. Photochem Photobiol. 1988;48(5):675–81.
39. Gomer CJ, Rucker N, Murphree AL. Differential cell photosensitivity following porphyrin photodynamic therapy. Cancer Res. 1988;48(16):4539–42.
40. West CM, West DC, Kumar S, Moore JV. A comparison of the sensitivity to photodynamic treatment of endothelial and tumour cells in different proliferative states. Int J Radiat Biol. 1990;58(1):145–56.
41. Fingar VH, Wieman TJ, Wiehle SA, Cerrito PB. The role of microvascular damage in photodynamic therapy: the effect of treatment on vessel constriction, permeability, and leukocyte adhesion. Cancer Res. 1992;52(18):4914–21.
42. Fingar VH, Kik PK, Haydon PS, Cerrito PB, Tseng M, Abang E, et al. Analysis of acute vascular damage after photodynamic therapy using benzoporphyrin derivative (BPD). Br J Cancer. 1999;79(11–12):1702–8.
43. He C, Agharkar P, Chen B. Intravital microscopic analysis of vascular perfusion and macromolecule extravasation after photodynamic vascular targeting therapy. Pharm Res. 2008;25(8):1873–80.
44. Debefve E, Cheng C, Schaefer SC, Yan H, Ballini JP, van den Bergh H, et al. Photodynamic therapy induces selective extravasation of macromolecules: insights using intravital microscopy. J Photochem Photobiol B, Biol. 2010;98(1):69–76.
45. Khurana M, Moriyama EH, Mariampillai A, Wilson BC. Intravital high-resolution optical imaging of individual vessel response to photodynamic treatment. J Biomed Opt. 2008;13(4):040502.
46. Madar-Balakirski N, Tempel-Brami C, Kalchenko V, Brenner O, Varon D, Scherz A, et al. Permanent occlusion of feeding arteries and draining veins in solid mouse tumors by vascular targeted photodynamic therapy (VTP) with Tookad. PLoS One. 2010;5(4):e10282.

47. Dolmans DE, Kadambi A, Hill JS, Waters CA, Robinson BC, Walker JP, et al. Vascular accumulation of a novel photosensitizer, MV6401, causes selective thrombosis in tumor vessels after photodynamic therapy. Cancer Res. 2002;62(7):2151–6.
48. Dougherty TJ, Gomer CJ, Henderson BW, Jori G, Kessel D, Korbelik M, et al. Photodynamic therapy. J Natl Cancer Inst. 1998;90(12):889–905.
49. Korbelik M. PDT-associated host response and its role in the therapy outcome. Lasers Surg Med. 2006;38(5):500–8.
50. Garg AD, Galluzzi L, Apetoh L, Baert T, Birge RB, Bravo-San Pedro JM, et al. Molecular and translational classifications of DAMPs in immunogenic cell death. Front Immunol. 2015;6:588.
51. Magna M, Pisetsky DS. The alarmin properties of DNA and DNA-associated nuclear proteins. Clin Ther. 2016;38(5):1029–41.
52. Bhargava A, Mishra D, Banerjee S, Mishra PK. Dendritic cell engineering for tumor immunotherapy: from biology to clinical translation. Immunotherapy. 2012;4(7):703–18.
53. Henderson BW, Gollnick SO. Mechanistic principles of photodynamic therapy. In: Vo-Dinh T, editor. Biomedical photonics handbook. Boca Raton: CRC Press; 2003. p. 36.1–27.
54. Oleinick NL, Evans HH. The photobiology of photodynamic therapy: cellular targets and mechanisms. Radiat Res. 1998;150(5 Suppl):S146–56.
55. Gollnick SO, Owczarczak B, Maier P. Photodynamic therapy and anti-tumor immunity. Lasers Surg Med. 2006;38(5):509–15.
56. Korbelik M, Sun J, Cecic I. Photodynamic therapy-induced cell surface expression and release of heat shock proteins: relevance for tumor response. Cancer Res. 2005;65(3):1018–26.
57. Korbelik M, Stott B, Sun J. Photodynamic therapy-generated vaccines: relevance of tumour cell death expression. Br J Cancer. 2007;97(10):1381–7.
58. Vabulas RM, Wagner H, Schild H. Heat shock proteins as ligands of toll-like receptors. Curr Top Microbiol Immunol. 2002;270:169–84.
59. Gomer CJ, Ryter SW, Ferrario A, Rucker N, Wong S, Fisher AM. Photodynamic therapy-mediated oxidative stress can induce expression of heat shock proteins. Cancer Res. 1996;56(10):2355–60.
60. Gollnick SO, Kabingu E, Kousis PC, Henderson BW. Stimulation of the host immune response by photodynamic therapy (PDT). Proc SPIE. 2004;5319:60–70.
61. Stott B, Korbelik M. Activation of complement C3, C5, and C9 genes in tumors treated by photodynamic therapy. Cancer Immunol Immunother. 2007;56(5):649–58.
62. Krosl G, Korbelik M, Dougherty GJ. Induction of immune cell infiltration into murine SCCVII tumour by photofrin-based photodynamic therapy. Br J Cancer. 1995;71(3):549–55.
63. Korbelik M, Cecic I. Contribution of myeloid and lymphoid host cells to the curative outcome of mouse sarcoma treatment by photodynamic therapy. Cancer Lett. 1999;137(1):91–8.
64. de Vree WJ, Essers MC, Koster JF, Sluiter W. Role of interleukin 1 and granulocyte colony-stimulating factor in photofrin-based photodynamic therapy of rat rhabdomyosarcoma tumors. Cancer Res. 1997;57(13):2555–8.
65. Kousis PC, Henderson BW, Maier PG, Gollnick SO. Photodynamic therapy enhancement of antitumor immunity is regulated by neutrophils. Cancer Res. 2007;67(21):10501–10.
66. Korbelik M, Cecic I. Mechanism of tumor destruction by photodynamic therapy. In: Nalwa HS, editor. Handbook of photochemistry and photobiology. Stevenson Ranch: American Scientific Publishers; 2003. p. 39–77.
67. Sun J, Cecic I, Parkins CS, Korbelik M. Neutrophils as inflammatory and immune effectors in photodynamic therapy-treated mouse SCCVII tumours. Photochem Photobiol Sci. 2002;1(9):690–5.
68. Gollnick SO, Evans SS, Baumann H, Owczarczak B, Maier P, Vaughan L, et al. Role of cytokines in photodynamic therapy-induced local and systemic inflammation. Br J Cancer. 2003;88(11):1772–9.
69. Hunt DW, Levy JG. Immunomodulatory aspects of photodynamic therapy. Expert Opin Investig Drugs. 1998;7(1):57–64.
70. Yusuf N, Katiyar SK, Elmets CA. The immunosuppressive effects of phthalocyanine photodynamic therapy in mice are mediated by CD4+ and CD8+ T cells and can be adoptively transferred to naive recipients. Photochem Photobiol. 2008;84(2):366–70.

71. Korbelik M, Krosl G, Krosl J, Dougherty GJ. The role of host lymphoid populations in the response of mouse EMT6 tumor to photodynamic therapy. Cancer Res. 1996;56(24):5647–52.
72. Canti GL, Lattuada D, Nicolin A, Taroni P, Valentini G, Cubeddu R. Immunopharmacology studies on photosensitizers used in photodynamic therapy. Proc SPIE. 1994;2078:268–75.
73. Korbelik M, Dougherty GJ. Photodynamic therapy-mediated immune response against subcutaneous mouse tumors. Cancer Res. 1999;59(8):1941–6.
74. Mroz P, Szokalska A, Wu MX, Hamblin MR. Photodynamic therapy of tumors can lead to development of systemic antigen-specific immune response. PLoS One. 2010;5(12):e15194.
75. Mroz P, Vatansever F, Muchowicz A, Hamblin MR. Photodynamic therapy of murine mastocytoma induces specific immune responses against the cancer/testis antigen P1A. Cancer Res. 2013;73(21):6462–70.
76. Maeurer MJ, Gollin SM, Storkus WJ, Swaney W, Karbach J, Martin D, et al. Tumor escape from immune recognition: loss of HLA-A2 melanoma cell surface expression is associated with a complex rearrangement of the short arm of chromosome 6. Clin Cancer Res. 1996;2(4):641–52.
77. Wachowska M, Gabrysiak M, Muchowicz A, Bednarek W, Barankiewicz J, Rygiel T, et al. 5-Aza-2′-deoxycytidine potentiates antitumour immune response induced by photodynamic therapy. Eur J Cancer. 2014;50(7):1370–81.
78. Abdel-Hady ES, Martin-Hirsch P, Duggan-Keen M, Stern PL, Moore JV, Corbitt G, et al. Immunological and viral factors associated with the response of vulval intraepithelial neoplasia to photodynamic therapy. Cancer Res. 2001;61(1):192–6.
79. Kabingu E, Vaughan L, Owczarczak B, Ramsey KD, Gollnick SO. CD8+ T cell-mediated control of distant tumours following local photodynamic therapy is independent of CD4+ T cells and dependent on natural killer cells. Br J Cancer. 2007;96(12):1839–48.
80. Henderson BW, Gollnick SO, Snyder JW, Busch TM, Kousis PC, Cheney RT, et al. Choice of oxygen-conserving treatment regimen determines the inflammatory response and outcome of photodynamic therapy of tumors. Cancer Res. 2004;64(6):2120–6.
81. Reis e Sousa C. Activation of dendritic cells: translating innate into adaptive immunity. Curr Opin Immunol. 2004;16(1):21–5.
82. Benvenuti F. The dendritic cell synapse: a life dedicated to T cell activation. Front Immunol. 2016;7:70.
83. Sur BW, Nguyen P, Sun CH, Tromberg BJ, Nelson EL. Immunophototherapy using PDT combined with rapid intratumoral dendritic cell injection. Photochem Photobiol. 2008;84(5):1257–64.
84. Castellino F, Germain RN. Cooperation between CD4+ and CD8+ T cells: when, where, and how. Annu Rev Immunol. 2006;24:519–40.
85. Gollnick SO, Vaughan L, Henderson BW. Generation of effective antitumor vaccines using photodynamic therapy. Cancer Res. 2002;62(6):1604–8.
86. Korbelik M, Sun J. Photodynamic therapy-generated vaccine for cancer therapy. Cancer Immunol Immunother. 2006;55(8):900–9.
87. Korbelik M, Merchant S, Huang N. Exploitation of immune response-eliciting properties of hypocrellin photosensitizer SL052-based photodynamic therapy for eradication of malignant tumors. Photochem Photobiol. 2009;85(6):1418–24.
88. Jalili A, Makowski M, Switaj T, Nowis D, Wilczynski GM, Wilczek E, et al. Effective photoimmunotherapy of murine colon carcinoma induced by the combination of photodynamic therapy and dendritic cells. Clin Cancer Res. 2004;10(13):4498–508.
89. Gomer CJ, Ferrario A, Murphree AL. The effect of localized porphyrin photodynamic therapy on the induction of tumour metastasis. Br J Cancer. 1987;56(1):27–32.
90. van Duijnhoven FH, Aalbers RI, Rovers JP, Terpstra OT, Kuppen PJ. Immunological aspects of photodynamic therapy of liver tumors in a rat model for colorectal cancer. Photochem Photobiol. 2003;78(3):235–40.

91. Dragieva G, Hafner J, Dummer R, Schmid-Grendelmeier P, Roos M, Prinz BM, et al. Topical photodynamic therapy in the treatment of actinic keratoses and Bowen's disease in transplant recipients. Transplantation. 2004;77(1):115–21.
92. Thong PS, Ong KW, Goh NS, Kho KW, Manivasager V, Bhuvaneswari R, et al. Photodynamic-therapy-activated immune response against distant untreated tumours in recurrent angiosarcoma. Lancet Oncol. 2007;8(10):950–2.
93. Thong PS, Olivo M, Kho KW, Bhuvaneswari R, Chin WW, Ong KW, et al. Immune response against angiosarcoma following lower fluence rate clinical photodynamic therapy. J Environ Pathol Toxicol Oncol. 2008;27(1):35–42.
94. Friedberg JS, Mick R, Stevenson JP, Zhu T, Busch TM, Shin D, et al. Phase II trial of pleural photodynamic therapy and surgery for patients with non-small-cell lung cancer with pleural spread. J Clin Oncol. 2004;22(11):2192–201.
95. Kabingu E, Oseroff AR, Wilding GE, Gollnick SO. Enhanced systemic immune reactivity to a basal cell carcinoma associated antigen following photodynamic therapy. Clin Cancer Res. 2009;15(13):4460–6.
96. Gad F, Zahra T, Francis KP, Hasan T, Hamblin MR. Targeted photodynamic therapy of established soft-tissue infections in mice. Photochem Photobiol Sci. 2004;3(5):451–8.
97. Tanaka M, Mroz P, Dai T, Huang L, Morimoto Y, Kinoshita M, et al. Photodynamic therapy can induce a protective innate immune response against murine bacterial arthritis via neutrophil accumulation. PLoS One. 2012;7(6):e39823.
98. Gryson O. Servier medical art France: servier; 2016 [Available from: http://www.servier.com/Powerpoint-image-bank.

Chapter 7
Multimodality Dosimetry

Martha Simões Ribeiro and Caetano Padial Sabino

Abstract PDT requires a multimodality approach for dosimetry because it works based on three essential components: light, photosensitizer, and molecular oxygen. Since these components are found in variable amounts inside target cells, PDT dosimetry is rather intricate. This chapter intends to address, with little mathematical complexity, the physical and chemical quantities that are most relevant for light and photosensitizer dosimetry as well as to present basic aspects of oxygen supply to achieve successful PDT interventions.

7.1 Introduction

To discuss dosimetry in PDT, it is first necessary to define the chemical and physical quantities associated to it. It is useful to do analogies with the dosimetry of ionizing radiation that is employed for radioprotection and radiotherapy purposes. In radiotherapy, the dose of radiation absorbed is expressed in gray units (i.e., 1 Gy = 1 J/kg). The absorbed dose in Gy is calculated as the energy delivered per kilogram of patient body mass. For phototherapies it is not as easy to use absorbed dose per weight as a

---

M.S. Ribeiro (✉)
Center for Lasers and Applications, Nuclear and Energy Research Institute, National Commission for Nuclear Energy, Av. Lineu Prestes 2242, Cidade Universitária, Sao Paulo 05508-000, SP, Brazil
e-mail: marthasr@usp.br

C.P. Sabino
Department of Microbiology, Institute for Biomedical Sciences, University of São Paulo, Av. Lineu Prestes 1347, Cidade Universitária, Sao Paulo 05508-000, SP, Brazil

Department of Clinical Analysis, School of Pharmaceutical Sciences, University of São Paulo, Sao Paulo, Brazil

Center for Lasers and Applications, Nuclear and Energy Research Institute, National Commission for Nuclear Energy, Sao Paulo, SP, Brazil
e-mail: caetanosabino@gmail.com

F.P. Sellera et al. (eds.), *Photodynamic Therapy in Veterinary Medicine: From Basics to Clinical Practice*, DOI 10.1007/978-3-319-45007-0_7

direct correlation to the associated biological effects since light is much more intensely attenuated by biological tissues and can be very influenced by several variables. To pursue this ideal situation, we must also consider some relevant optical properties of tissues, such as target depth and rates of light scattering and absorption, to calculate the energy adequate to the treatment. Moreover, light-absorbing pigments are present in different concentrations among tissues, and their exact concentration and distribution in tissues can present great variations among different individuals.

To properly establish the dosimetry for PDT is quite challenging since many factors can influence the efficacy of the treatment (e.g., heterogeneous tissue optics, variable light absorption due to blood content, variable oxygen tissue levels, etc.). In regard to light parameters, one can obtain very different results if the same energy density is used for variable power density and exposure time [1]. In addition, killing efficiency is also directly dependent on target concentration of photosensitizer (PS) and molecular oxygen. Hence, to resume PDT dosimetry only in respect to the amount of light delivered is often faulty because it does not consider individual variability of PS uptake by different cell types, oxygenation and optical properties of tissues, presence of body fluids, or, consequently, interdependence of these factors (PS, light, and tissue). This chapter will cover the basic aspects of PDT dosimetry to facilitate the interpretation on which parameters should be examined to obtain effective PDT outcomes.

7.2 Light

7.2.1 Light Sources

Light radiation used in PDT represents a tiny fraction of the electromagnetic spectrum (see Chap. 2). Due to its different types of interaction with matter, light is characterized into different sections of the electromagnetic spectrum: the ultraviolet (UV) light lays in between 100 and 400 nm, visible light from 400 to 700 nm, and infrared (IR) from 700 to 100,000 nm. Light behaves either as waves or particles (wave-particle duality). The wave characteristic normally is more relevant when light propagation through space is analyzed. On the other hand, when we analyze light interactions with atoms or molecules, its particle or quantum nature is often more pertinent.

Light devices are the major effectors of PDT. Because of the wide range of possible applications of light for diagnosis or therapy remains expanding, the demand on new light sources has been intensely increasing over the last few years. Hence, manufacturers are constantly developing several innovative types of light delivery systems for those devices that may present optimized characteristics for some specific applications. Even though, the most important aspects of light sources rely on the produced light. In general, current protocols require to use monochromatic light sources to achieve its best efficiency. Among them, most available technologies obtain monochromatic light from lasers, light-emitting diodes (LEDs), or filtered light from broad-spectrum incandescent lamps. We will describe below some of the main characteristics of light sources currently used for PDT.

7.2.1.1 Lasers

The first light amplification by stimulated emission radiation (LASER) device was developed by Theodore Maiman in 1960. A device requires three basic components to generate laser radiation: (1) an active medium, i.e., a collection of atoms, molecules, or ions that emit radiation; (2) an energy source to excite the active media capable to induce population inversion, i.e., when components of active media are more prevalently in the excited state than in ground state, turning the active medium into an amplifier of radiation; and (3) a resonant cavity to reflect produced light back into the system (e.g., a pair of mirrors facing each other) and induce the stimulated emission of radiation.

Lasers are distinguished from other conventional light sources due to some of their unique properties such as monochromaticity (i.e., emission of a very narrow spectral band of radiation, generally presenting a bandwidth in the order of 2 nm or less), optical fiber coupling, coherence (i.e., emitted electromagnetic waves are in spatial and temporal phase in relation to each other), and collimation (i.e., light travels through long distances with very low divergence).

7.2.1.2 LED

LEDs were first built in 1962 by Nick Holonyak Jr. and currently represent a class of very efficient electroluminescent devices capable of emitting relatively monochromatic light. LEDs are basically semiconductors that emit light when an electric current flows through them in a certain direction. Light emitted from LED is considered to be semi-monochromatic because even though it consists of a narrow spectral band (e.g., emission bandwidth around 50 nm), lasers can emit narrower bands. The color or central wavelength of emitted light (i.e., energy of emitted photons) is determined by the specific energy of the semiconductor bandgap. When a free electron is captured to an orbital of the semiconductor bandgap, the energy difference of the free electron and the captured electron is released in the form of heat or light.

7.2.2 Light-Tissue Interaction

Light can either be reflected, absorbed, scattered, or transmitted from biological tissues or any other material media (Fig. 7.1). Reflection occurs when light encounters a physical interface between two materials of different refractive indexes, such as among air and water or a biological tissue. If the surface is smooth, it is assumed that its irregularities are small compared to the wavelength of incident radiation and results in specular reflection, i.e., similar to what happens in mirrors. On the other hand, diffuse reflection predominates if the surface roughness is composed of irregularities spaced by equal to or larger distances than the wavelength magnitude of the incident radiation, i.e., similar to light reflection from a white paper sheet. The latter case is the most common phenomenon to biological tissues. At a 90° incidence on

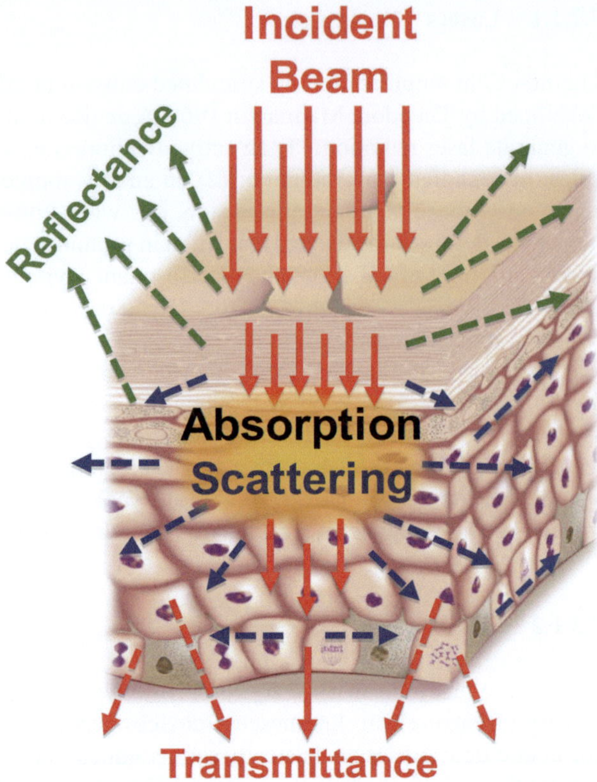

Fig. 7.1 Schematic illustration representing light interactions with biological tissues (e.g., skin). When incident light penetrates a tissue, it can be reflected, absorbed, scattered, or transmitted

skin, there is a loss of 4–7 % of light due to reflection because of divergences between the refractive indexes of the air-stratum corneum interface ($n_{air} = 1$ and $n_{sc} = 1.55$) [2].

The remaining fraction of radiation is transmitted to the deeper tissue layers and can then be absorbed or scattered by biomolecules or photosensitizers. These two processes determine the depth of light penetration into the tissue as well as the amount of light that will be remitted from the tissue due to backscattering events. These two main processes of interaction between light and tissue for PDT will be further detailed in the next sections.

7.2.2.1 Absorption

Light absorption occurs when electromagnetic radiation does not return to the incident surface neither propagates into the medium. Absorption occurs due to a transfer of light energy to an electron present in a molecule or atom, which promotes the atom to a higher energy state that can posteriorly decay back to its original configuration dissipating the energy absorbed from light (e.g., chemical, thermal, or kinetic processes). Hence, light intensity can be attenuated when transmitted through a

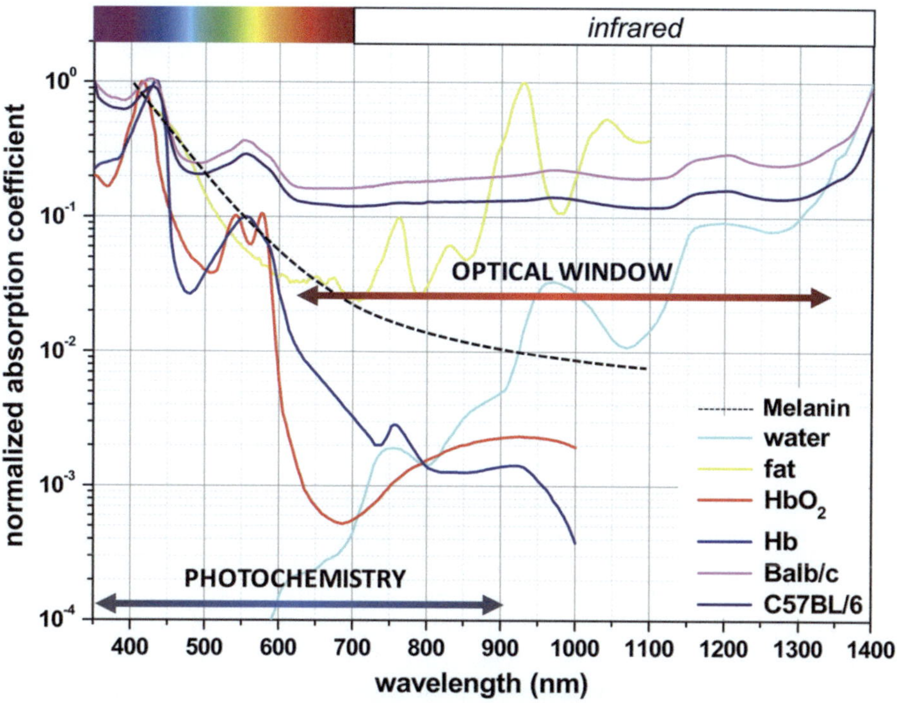

Fig. 7.2 Attenuation spectra of albino (Balb/c) and pigmented (C57BL/6) mouse skin together with absorption spectra of the most important chromophores present in biological tissues: melanin, water, fat, hemoglobin (Hb), and oxyhemoglobin (HbO_2). Observe that there is a region of minimum light-tissue interaction between 600 and 1300 nm, commonly referred in literature as the "optical window"

material medium. The rate of light attenuation in biological tissues is governed by the sum of absorption and scattering coefficients (in cm^{-1}), which are directly proportional to the probability of absorption or scattering events during light propagation.

A transparent medium allows light to pass free of absorption or scattering events, i.e., all light that enters the medium is transmitted through without attenuation.

Each molecule has a characteristic absorption spectrum. In biological tissues, the main chromophores (i.e., molecules that absorb light) for visible and infrared light spectral regions are water, heme-containing proteins (e.g., hemoglobin, myoglobin, bilirubin, and cytochromes), melanin, and fat (Fig. 7.2).

7.2.2.2 Elastic Scattering

Light is elastically scattered when its propagation path is deflected by particles present inside the tissue without any changes in photon energy (see Fig. 7.1). The mechanism by which light is scattered mostly depends on the size of the scattering particle: those with dimensions much smaller than light wavelength tend to produce a more radially diffuse type of scattering referred as Rayleigh scattering, and those with dimensions comparable or larger than wavelength tend to produce a more forward-directed scattering, referred as Mie scattering.

Rayleigh scattering occurs when a particle absorbs a photon of certain energy and emits another photon with the same energy but in a random direction. This type of scattering has a high dependence on light wavelength, i.e., the scattering intensity is inversely proportional to the fourth power of photon wavelength. Hence, blue photons are much more scattered than red ones. This is the phenomenon that explains the blue color of the sky. On the other hand, Mie scattering occurs mainly due to changes in photon trajectory due to reflection or refraction in large particles. It shows a much weaker dependence on the wavelength as the scattering intensity is inversely proportional to the fourth root of photon wavelength. Hence, blue photons are scattered slightly more than red photons. Conversely, this phenomenon is what determines the white color of clouds as light is scattered by relatively large waterdrops (Fig. 7.3).

In complex media such as biological tissues, both types of scattering phenomena occur simultaneously. Small particles such as macromolecules mostly produce Rayleigh scattering while larger particles such as cells and organelles produce Mie scattering. To combine both scattering events to a single parameter, it is necessary to define a probability function $p(\theta)$ of photon scattering in relation to an angle θ. If the probability $p(\theta)$ presents little dependence on θ, the scattering is termed as isotropic (i.e., light is equally scattered to all angles). Otherwise, the scattering is termed anisotropic (see Fig. 7.3). The rate of anisotropy is given by anisotropy coefficient g, where $g=1$ results in no scattering probability, $g=-1$ means that all light is backscattered, and $g=0$ indicates completely isotropic scattering. By definition, the anisotropy coefficient is the average cosine of the scattering angle θ. For most biological tissues, g can assume values between 0.7 and 0.99. Therefore, the correspondent scattering angles are between 8 and 45°.

Putting together all influence of scattering and absorbing components of tissues, we can obtain the resultant attenuation coefficient. To privilege light penetration into tissues, we must avoid excessive attenuation either due to absorption by tissue chromophores or scattering by tissue particles. In fact, there is an optical window for light penetration into tissues (i.e., where absorption and scattering are minima) in between 600 and 1000 nm. For this reason, PDT applications targeting cells deeply localized in tissues (>1 mm) usually use strategies employing photosensitizers that absorb light in this spectral range. The major explanations for such privileged light transmission in this wavelength range rely on two major factors:

1. The predominant interaction is forward scattering, that is, the light mostly scatters in the direction of beam propagation.
2. The absorption of the light radiation in this range of wavelengths by water and blood, the main chromophores of the biological tissue, is small, and then the beam transmission is favored allowing a more significant interaction with deeper tissues (see Fig. 7.2).

For antimicrobial PDT, such importance of light penetration into tissue is often not so relevant because the infected anatomical region is usually superficial (<1 mm

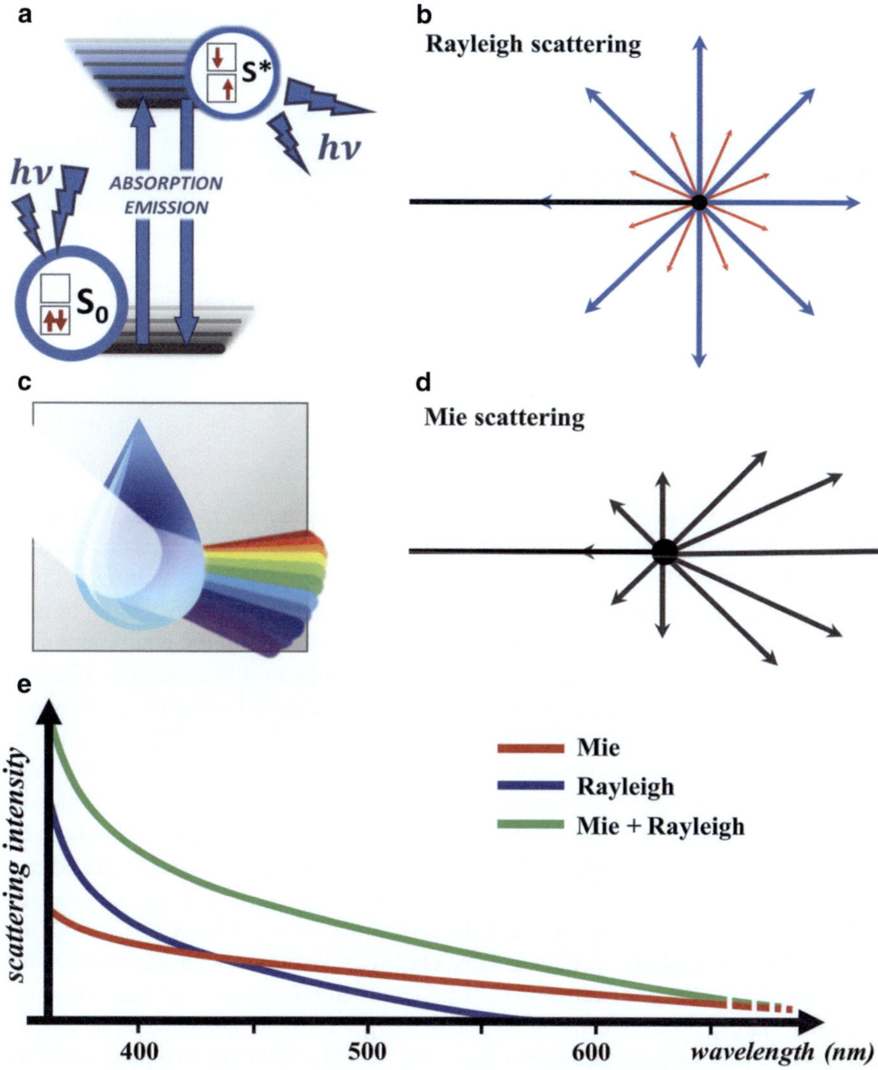

Fig. 7.3 Illustrative description of light scattering phenomena. Rayleigh scattering occurs when very small particles (i.e., size $<<\lambda$) absorb and emit photons of same energy (**a**) isotropically (**b**). On the other hand, Mie scattering occurs when larger particles (i.e., size $\geq\lambda$) deflect photon trajectories by refraction or reflection (**c**). In this case, forward scattering is more privileged (**d**). While Rayleigh scattering presents high dependence on wavelength ($\propto\lambda^{-4}$), Mie scattering intensity is only weakly influenced by wavelength ($\propto \sqrt[4]{\lambda}$). In biological tissues both types of light scattering occur simultaneously (**e**)

deep) and localized. For this sort of PDT application, wavelengths between 400 and 600 nm have been successfully used to treat in vivo infected lesions. Note that Niels Finsen was awarded with a Nobel Prize in 1903 for the successful use of filtered sunlight to treat cutaneous tuberculosis (see Chap. 1). Recent literature also reports that a violet-blue light (between 400 and 420 nm) can even rescue mice from lethal infections by inactivating pathogens without the need to introduce exogenous photosensitizers [3, 4]. In this approach, violet-blue light directly acts on endogenous bacterial chromophores, such as derivatives of porphyrins, to selectively inactivate pathogen cells that tend to accumulate it due to their accelerated metabolism. Even though effective, this approach requires very long light exposures because such endogenous photosensitizers are present in relatively low concentrations and adding an exogenous photosensitizer can greatly enhance PDT efficiency and effectiveness [5]. It is important to mention that white lamps and sunlight have also been used to inactivate different microorganisms [6, 7].

7.2.3 *Important Light-Related Parameters*

7.2.3.1 Wavelength

As presented in Chap. 2, the wavelength of electromagnetic radiation is directly proportional to the energy that each individual photon carries. Photosensitizers only absorb light at wavelengths correspondent to the specific energy required to promote one of its outermost electrons to an excited state. So when a clinician uses some particular photosensitizer for PDT, the light emitted from the equipment used for irradiation must overlap with the photosensitizer absorption spectrum. The best efficiency to induce photodynamic reactions is only achieved when the peak wavelength of light source emission and photosensitizer absorption are perfectly matched. For example, methylene blue has its peak absorption at 664 nm; hence, the most efficient light source combination would be a 664 nm laser due to its very narrow emission band centered at the same wavelength that is most intensely absorbed by methylene blue. Some LED-based light sources, also with peak emission at 664 nm, may bring similar results with methylene blue as well, even though it would require slightly longer exposure times to reach the same final results achieved by irradiation performed with laser light, under the same energy and irradiance parameters. On the other hand, LEDs have the advantage of being much more cost- and time-effective to irradiate larger areas because the spot size of LED-based equipments is usually much greater and homogeneous than those produced by laser-based equipments. An emission spectrum of hypothetical laser- and LED-based light sources can be observed in Fig. 7.4.

7.2.3.2 Power and Energy

The average optical power of light sources is one of the most important light parameters required to obtain consistently positive results when performing PDT. Based on this parameter and the light exposure time, we calculate the total amount of light energy delivered during a PDT irradiation procedure. All LED-based light sources and most lasers commercially available operate under a continuous light emission

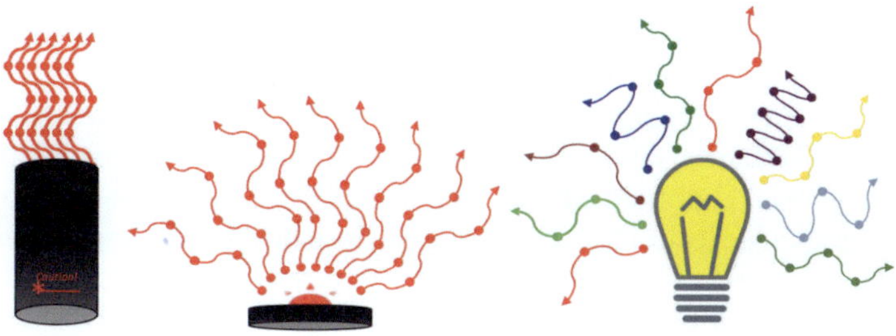

	LASER	LED	white light
monochromatic	✓✓	✓	✓ (+ filters)
collimated	✓✓	✓ (+ lens)	✓ (+ lens)
coherent	✓✓	✗	✗
optical fiber	✓✓	✓ (+ lens)	✓ (+ lens)
power efficiency	✓✓	✓✓	✗
cost	✗	✓✓	✓

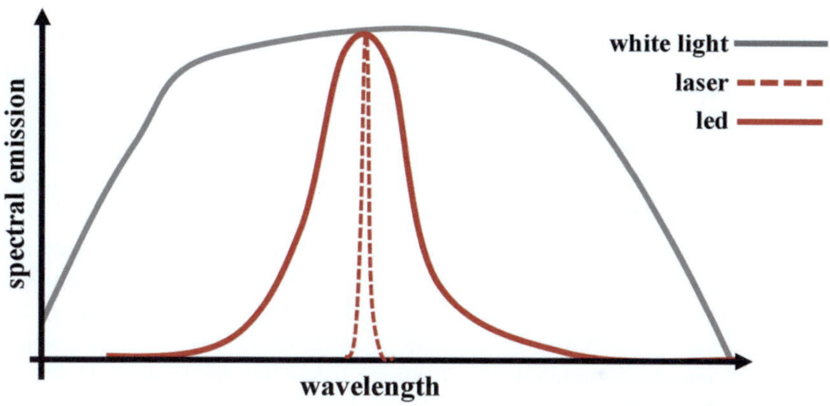

Fig. 7.4 Schematic illustration of the main technical divergences found among the most commonly used light sources. Lasers and LED can efficiently produce monochromatic light without the need for a filter. Yet, lasers emit collimated and coherent light (i.e., photons propagate in phase toward a single direction) that can be easily coupled to an optical fiber. LED and light bulbs produce diffuse light that needs to be converged by lens systems to be focused onto a light spot or into optical fibers. Below emission bands of white light, laser, and LED can be observed

regimen. However, when the light source is a pulsed laser, the optical power varies between a maximum value (peak power) and zero. In this case, the most relevant parameter for calculation of PDT dosimetry is the mean power emitted by the laser over time. If the light source emits continuously, the optical power remains constant over time and is equal to the mean power. The expression that relates power and energy is

$$E[J] = P[W] \times t[s]$$

where E is the energy delivered given in joules (J), P is the mean optical power in watts (W), and t is the exposure time in seconds (s). A commercially acquired equipment should inform the typical output power in user's manual. However, since light sources and electronic components can lose their efficiency over time, it is advisable to have a proper optical power meter to measure the actual output power regularly and prevent from using it with incorrect parameters. This attention should especially be taken when coupling an optical fiber to laser tip. In some occasions fibers may have imperfections along them, causing important variations on output optical power as will be described over the next section.

It is worthy to mention that some equipments use white light or white light associated with filters to provide monochromatic light for PDT. In this case, a radiometer is recommended to measure the actual spectral emission of this particular light source.

7.2.3.3 Power Density (Irradiance or Fluence Rate)

An appropriate definition must be established to correctly understand the term power density. The American Association of Physicists in Medicine (AAPM) defines the following parameters [8]:

- Radiant power: total optical power emitted by a source, transferred or received as electromagnetic radiation. The unit used according to the International System (SI) is the watt (W).
- Irradiance (power density): optical power incident on a surface divided by the area covered by light spot. In SI, the unit used is W/m^2, but in phototherapy field mW/cm^2 is more commonly used. 1 W = 1000 mW; 1 m^2 = 10,000 cm^2.
- Fluence rate: optical power incident on a surface divided by the area covered by light spot plus backscattered light from the tissue into the surface. Hence, light that interacted with target tissue is also taken in account. Also in this case, the SI unit is W/m^2 or mW/cm^2 in accordance with phototherapy community.

It is noteworthy that in most articles found in the literature, the authors use the term intensity instead of irradiance. In fact, the International Commission on Illumination defines "intensity" as the quotient between radiant power and the surface area at a solid angle. A solid angle in steradian (sr) equals the area of a segment of a unit sphere in the same way a planar angle in radian equals the length of an arc of a unit circle. Therefore, the unity of light intensity in SI is W/sr. This quantity considers the propagation of light after penetrating the tissue in a given direction. Unfortunately, even with all current technology, it is still difficult to measure light intensity as described by its definition [9].

Observe that irradiance and fluence rate have the same unit (power per area), but they do not represent the same physical quantity. The irradiance (power density) is defined for the light incident on tissue surface, while the fluence rate is defined for the light that penetrated into the irradiated tissue and was backscattered to its surface.

To calculate irradiance, we use the following expression:

$$I\left[\frac{W}{\text{cm}^2}\right]=\frac{P[W]}{A\left[\text{cm}^2\right]}$$

Irradiance can indicate possibility of thermal effect. For example, consider a red laser with 100 mW coupled to an optical fiber diameter of 200 μm. The transection area of the fiber is 0.000314 cm^2 ($A=\pi r^2=3.14\times 100\times 10^{-8}$ cm^2). Assuming a good coupling so that the power after the fiber is 80 mW and using the above expression, it means that we have an irradiance about 255 W/cm^2. A reasonable estimative is that irradiances above 1 W/cm^2 may lead to thermal effects that can be sensed by the patient after a few seconds. If the fiber remains rested for a long time on the treatment site, there is a risk of temperature rise and consequent thermal injury to the tissue. It is also important emphasize that PDT does not rely on thermal effects to inactivate cells; cells are killed by intense oxidative stress induced by photochemical production of reactive oxygen species (ROS). Hence, the specificity of PDT interventions requires avoidance of excessive thermal effects.

7.2.3.4 Energy Density (Radiant Exposure or Fluence)

The energy density is the quantity commonly referred as light dose for PDT. This is the physical quantity that literature suggests as the most appropriate parameter to evaluate the possibility of the therapeutic effect. However, maintaining the same energy density but varying the power density and exposure time may bring divergent results [1].

Analogous to power density, energy density is the amount of energy delivered divided by the spot area on target surface. It is usually measured in J/cm^2. However, just as in the previous section, the AAPM defines other similar parameters used in phototherapies that need to be well defined [8]:

- Radiant exposure: the radiant energy incident on the surface containing the point of interest, for a period of time, per area. In SI, the unit used is J/m^2, but in phototherapy field J/cm^2 is more commonly used.
- Fluence: total optical energy delivered to a surface divided by the area covered by light spot plus backscattered light energy from the tissue into the surface. As the fluence rate, this quantity considers the interaction of light with the biological target. The unit in SI is also J/m^2.

The radiant exposure is calculated as follows:

$$D\left[\frac{J}{\text{cm}^2}\right]=\frac{E}{A}=\frac{P\times t}{A}$$

where D is the radiant exposure (i.e., dose), P is the mean power, t is the exposure time, and A is the area covered by light spot. Reformulating the expression above, we can obtain the exposure time,

$$t = \frac{D \times A}{P}$$

For example, if the dose used to treat an infected wound is 68.5 J/cm^2 using a light source with P=40 mW and beam diameter of 3 mm (radius = 1.5 mm or 0.15 cm), to calculate the exposure time, we first calculate the beam area ($A = \pi r^2 = 3.14 \times 0.15 = 0.07$ cm^2). Therefore, the exposure time will be

$$t = 0.07 \times 68.5 / 0.04 = 120\,\text{s or } 2\,\text{min}.$$

In brief, for light dosimetry we most commonly use irradiance and radiant exposure to describe the parameters of light. The irradiance gives the amount of optical power over a surface area, regardless of the direction of propagation, which may be absorbed or scattered. Therefore, the light dose is usually prescribed as the resultant radiant exposure that does not take in account reflected or scattered light.

7.3 Photosensitizer

7.3.1 Concentration

For photosensitizer dosimetry is essential to know the molecular concentration to be used. In chemistry, the unit used for concentration is molar (M), which is 1 mol/L (6.02×10^{23} molecules per liter). Many authors also use the PS concentration expressed in percentage of mass/volume. With a simple mathematical formula, we can relate percent and molar concentrations.

As an example, we will use the methylene blue (MB), which has an approximate molecular weight of 320 g. Therefore, 1 mol of MB is 320 g, and 1 M equals to 320 g/L. Consider a concentration of 0.01 %, i.e., 0.1 g of MB diluted in 1 L of water. So with a simple rule of three, we have the following formula:

$$C[M] = \frac{\text{sample mass}[\text{g}]}{\text{molar mass}[\text{g}]} = \frac{0.1\,\text{g}}{320\,\text{g}} \cong 0.0003\,\text{M} = 300\,\mu\text{M}$$

There are several new photosensitizer molecules under development for PDT use, as described in Chap. 3. In general, the ideal photosensitizer for PDT must have the following characteristics [10]:

1. Photophysical: high-absorption (high-molar extinction coefficient, cm^{-1}M^{-1}) resonant to the wavelength provided by light source equipment.

2. Photochemical: long lifetime in the triplet state to allow molecular interaction with oxygen, high ROS yield (especially for singlet oxygen).
3. Chemical: high photostability, easy synthesis and purification, low cost, and be amphiphilic but with high solubility in water to allow systemic administration without a vehicle such as liposomes or emulsions (for topical application, this is not strictly necessary).
4. Biological: low toxicity in the dark, selective uptake in target tissue or cell, high affinity for subcellular targets that trigger cell death (e.g., mitochondria, endoplasmic reticulum, lysosomes).

Some photosensitizers, such as porphyrins, may present low photostability and suffer photodegradation (or photobleaching) after a period of light exposure [11]. This happens because photoexcitation may induce molecular cleavage or ROS may react with photosensitizer molecules and degrade them. So after some minutes inducing photosensitizer degradation, the remaining photoactive molecules may be in too small quantities to produce appreciable photodynamic cellular killing.

In clinical application of PDT, a key step is to decide which photosensitizer and which concentration should be used. Excessively high PS concentrations tend to form aggregates that usually present much reduced photodynamic activity and also shift its absorption peak reducing the efficiency of light absorption. This is particularly evident for planar molecules such as methylene blue [12]. In the presence of organic fluids (e.g., saliva or blood), photosensitizer tends to be further diluted and biomolecules from the fluid compete with target cells for interaction with photosensitizer, which may change its absorbance (Fig. 7.5).

Another important factor is that depending on the concentration of the PS, it can become an optical barrier (filter effect) so that light cannot penetrate into tissue to reach deeper target cells/tissues. Figure 7.6 illustrates this situation. Note that increasing methylene blue concentration, red light transmission through the solution is rapidly diminished.

7.3.2 Pre-irradiation Time

The pre-irradiation time (PIT) is defined as the time interval in which photosensitizer is in contact with target tissue in the dark to allow saturation of cellular uptake, i.e., drug-light administration interval. This contact time is necessary for the photosensitizer diffusion through the organisms (in case of systemic administration), tissues, and cells. Literature reports that the effects of the PDT are dependent on the PIT used [13, 14], ranging from minutes to days depending on treatment modality. In addition, some photosensitizers may present intrinsic cytotoxicity (especially high concentrations for long periods) that may inactivate microorganisms or tumor cells even in the absence of light exposure.

For antimicrobial PDT, the PIT may affect the results dependent on the microorganism. For example, Jackson et al. investigated the effect of PDT in

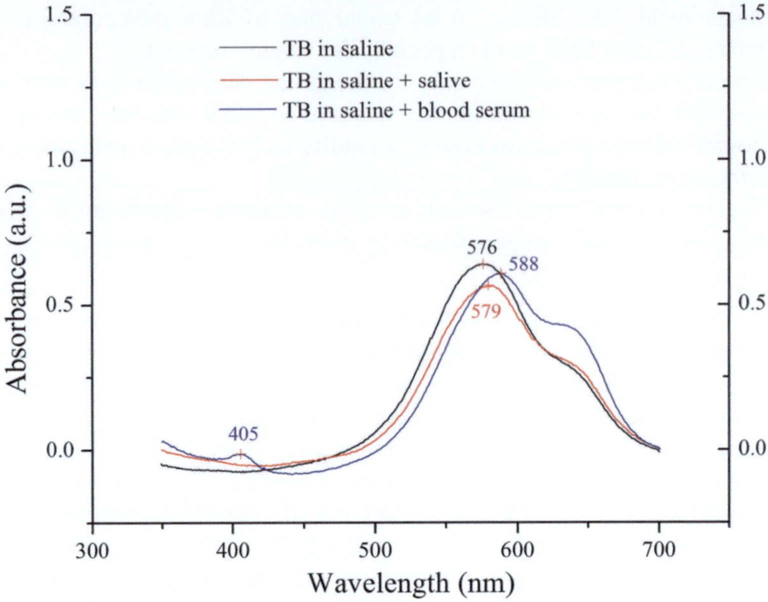

Fig. 7.5 Absorbance of toluidine blue (TB) diluted in saline, saline + saliva, or saline + blood serum solution. Observe that the maximum absorbance peak shifts in the presence of organic fluids. Particularly for blood, a new band at 405 nm also can be noticed due to blood-derived porphyrins (spectra were kindly provided by Silvia Cristina Núñez)

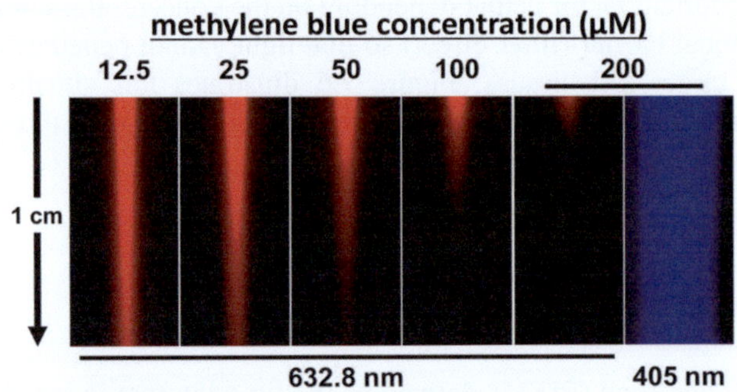

Fig. 7.6 Transmission of red laser light through different concentrations of methylene blue (MB) solutions. Observe that as MB concentration is increased, less light is transmitted. On the other hand, blue light is totally transmitted even through the highest concentration of MB solution tested

Candida albicans yeast forms and hyphae, to determine optimal conditions of inactivation. The best results were energy dependent (42 J was optimal for both forms); the optimal concentrations of PS were 25 mg/mL and 12.5 mg/mL for yeast and hyphae, respectively; the optimal PIT was 5 min for yeast, while for the form of hyphae, results did not depend on PIT [15].

Table 7.1 Characteristics of light source and photosensitizer that must be considered to perform PDT

Light source	Photosensitizer
Wavelength (nm)	Molar mass (g/mol)
Irradiance, fluence rate, or power density (W/cm^2)	High quantum yield to produce ROS
Radiant exposure, fluence, or energy density (J/cm^2)	Molar extinction coefficient (cm^{-1}M^{-1})
Exposure time (s, min, h)	Concentration (mol/L)
	Pre-irradiation time (s, min, h)

Kömerik and coworkers aimed to determine the antimicrobial effect of toluidine blue (TB) and TB biodistribution through the buccal mucosa of rats by fluorescence microscopy. The authors reported that fluorescence was detected with a high TB concentration (200 μg/mL), particularly in the keratin layer of the epithelium, with almost no fluorescence in the connective tissue. Moreover, when left for 1 min, the PS was located in the keratinized layer. Increasing the PIT for 10 min, the TB penetration was more profound (2.5×) across the epithelial layer. The thickness of the keratinized epithelium also interfered with the penetration of the PS [16].

Table 7.1 summarizes the main features related to PDT dosimetry regarding light and PS.

7.4 Oxygen Supply

For effective PDT outcomes, we must ensure that a well-based and defined dosimetry is used for each treatment. As seen in this chapter, dosimetry is mostly related to light and photosensitizer parameters. However, photodynamic reactions directly depend on the presence of oxygen to generate ROS (see Chap. 2). These are three necessary components that together generate ROS.

Regarding antimicrobial PDT, generally microorganisms have defenses against some of ROS formed. For the superoxide anion, they have enzymes such as superoxide dismutase and, for hydrogen peroxide, have the catalase and peroxidases. Carotenoids are the more effective quenchers of singlet oxygen and occur for all photosynthetic organisms but rarely in nonphotosynthetic bacteria and eukaryotes. For hydroxyl radical, only external antioxidants (e.g., glutathione peroxidase) would help in the fight against ROS generated during photodynamic effect.

To monitor ROS formed during PDT is not an easy task; their lifetime is very short (less than a millisecond) since they rapidly react with biomolecules. However, recent advances show that it is feasible to directly detect singlet oxygen luminescence (at 1270 nm) during PDT to allow a ROS-based dosimetry [17]. Indirect methods for studying ROS involved with photodynamic effect can also be used [18].

Cancer PDT studies show that fractionated light irradiation, i.e., when the irradiation is interrupted for a short period of time, increases the photodynamic effect by induction of reoxygenation of the tissue [19]. This strategy can be effective to attack hypoxic regions of solid tumors and highly inflamed areas but, despite the good results, a reoxygenation of tissue can also induce a faster photobleaching of photosensitizer [20]. Thus, readministration of photosensitizer may also be recommended.

The usual method for determination of PDT dosimetry is based on photosensitizer concentration, pre-irradiation time, and light parameters such as irradiance and radiant exposure. This approach is usually effective for superficial lesions but may be too simplistic for large solid tumors. If target tissues or cells have low oxygen supply, the yield of ROS will be diminished. In addition, PS concentration, light propagation, and tissue oxygenation can modify during PDT, and one parameter can influence the other. In this context, recent advances have been achieved to optimize PDT dosimetry via the use of sophisticated mathematical models of light propagation into tissues [21], monitoring of PS photobleaching [11, 22], and new technologies to measure singlet oxygen production [17]. PDT induces biological response that is correspondent to the amount of ROS produced (see Chap. 5). To solve such dosimetry intricacies, some experimental models are being developed as follows [10]:

1. Direct dosimetry, in which $^1O_2$ is measured.
2. Implicit dosimetry, in which a reporter for ROS production is measured, e.g., the direct measurement of PS photobleaching by fluorescence.
3. Explicit dosimetry, in which the three fundamental parameters (light, PS, and oxygen) are used to calculate $^1O_2$ through the most significant reaction pathways.

In summary, understanding light parameters is fundamental to obtain good results from PDT. The availability of oxygen in the microenvironment is also an important factor to consider. Moreover, attention should also be attained to photosensitizer administration in respect to the concentration, pre-irradiation time, light source used for excitation, and bioavailability in the target tissue. Thus, for a successful application of PDT in veterinary medicine, we must take into account all aspects related to dosimetry addressed in this chapter.

References

1. Prates RA, da Silva EG, Yamada Jr AM, Suzuki LC, Paula CR, Ribeiro MS. Light parameters influence cell viability in antifungal photodynamic therapy in a fluence and rate fluence-dependent manner. Laser Phys. 2009;19(5):1038–44.
2. Castronuovo G, Fava G, Giavelli S. The skin role during a low level laser therapy. Lasers applications in medicine. Bologna: Monduzzi; 1992. p. 19–24.
3. Dai T, Gupta A, Huang YY, Sherwood ME, Murray CK, Vrahas MS, et al. Blue light eliminates community-acquired methicillin-resistant Staphylococcus aureus in infected mouse skin abrasions. Photomed Laser Surg. 2013;31(11):531–8.
4. Zhang Y, Zhu Y, Gupta A, Huang Y, Murray CK, Vrahas MS, et al. Antimicrobial blue light therapy for multidrug-resistant Acinetobacter baumannii infection in a mouse burn model: implications for prophylaxis and treatment of combat-related wound infections. J Infect Dis. 2014;209(12):1963–71.

5. Arenas Y, Monro S, Shi G, Mandel A, McFarland S, Lilge L. Photodynamic inactivation of Staphylococcus aureus and methicillin-resistant Staphylococcus aureus with Ru(II)-based type I/type II photosensitizers. Photodiagnosis Photodyn Ther. 2013;10(4):615–25.
6. Wood S, Metcalf D, Devine D, Robinson C. Erythrosine is a potential photosensitizer for the photodynamic therapy of oral plaque biofilms. J Antimicrob Chemother. 2006;57(4):680–4.
7. Enk CD, Nasereddin A, Alper R, Dan-Goor M, Jaffe CL, Wulf HC. Cutaneous leishmaniasis responds to daylight-activated photodynamic therapy: proof of concept for a novel self-administered therapeutic modality. Br J Dermatol. 2015;172(5):1364–70.
8. Hetzel F, Patterson M, Preuss L, Wilson B. Recommended nomenclature for physical quantities in medical applications of light. In: (AAPM) AAOPiM, editor. American Institute of Physics; Woodbury, NY, USA; 1996.
9. Sliney DH, Illumination ICo. Radiometric quantities and units used in photobiology and photochemistry: recommendations of the Commission Internationale de L'Eclairage (International Commission on Illumination). Photochem Photobiol. 2007;83(2):425–32.
10. Wilson BC, Patterson MS. The physics, biophysics and technology of photodynamic therapy. Phys Med Biol. 2008;53(9):R61–109.
11. Boere IA, Robinson DJ, de Bruijn HS, Kluin J, Tilanus HW, Sterenborg HJ, et al. Protoporphyrin IX fluorescence photobleaching and the response of rat Barrett's esophagus following 5-aminolevulinic acid photodynamic therapy. Photochem Photobiol. 2006;82(6):1638–44.
12. Usacheva MN, Teichert MC, Biel MA. The role of the methylene blue and toluidine blue monomers and dimers in the photoinactivation of bacteria. J Photochem Photobiol B. 2003;71(1–3):87–98.
13. Andrade MC, Ribeiro AP, Dovigo LN, Brunetti IL, Giampaolo ET, Bagnato VS, et al. Effect of different pre-irradiation times on curcumin-mediated photodynamic therapy against planktonic cultures and biofilms of Candida spp. Arch Oral Biol. 2013;58(2):200–10.
14. Pogue BW, O'Hara JA, Goodwin IA, Wilmot CJ, Fournier GP, Akay AR, et al. Tumor PO(2) changes during photodynamic therapy depend upon photosensitizer type and time after injection. Comp Biochem Physiol A Mol Integr Physiol. 2002;132(1):177–84.
15. Jackson Z, Meghji S, Macrobert A, Henderson B, Wilson M. Killing of the yeast and hyphal forms of candida albicans using a light-activated antimicrobial agent. Lasers Med Sci. 1999;14(2):150–7.
16. Kömerik N, Curnow A, MacRobert AJ, Hopper C, Speight PM, Wilson M. Fluorescence biodistribution and photosensitising activity of toluidine blue o on rat buccal mucosa. Lasers Med Sci. 2002;17(2):86–92.
17. Li B, Lin L, Lin H, Wilson BC. Photosensitized singlet oxygen generation and detection: Recent advances and future perspectives in cancer photodynamic therapy. J Biophotonics. 2016. doi: 10.1002/jbio.201600055.
18. Sabbahi S, Alouini Z, Jemli M, Boudabbous A. The role of reactive oxygen species in Staphylococcus aureus photoinactivation by methylene blue. Water Sci Technol. 2008;58(5): 1047–54.
19. Curnow A, Haller JC, Bown SG. Oxygen monitoring during 5-aminolaevulinic acid induced photodynamic therapy in normal rat colon. Comparison of continuous and fractionated light regimes. J Photochem Photobiol B. 2000;58(2–3):149–55.
20. Pudroma X, Juzeniene A, Ma LW, Iani V, Moan J. Fluorescence photobleaching of ALA and ALA-heptyl ester induced protoporphyrin IX during photodynamic therapy of normal hairless mouse skin: a comparison of two light sources and different illumination schemes. J Environ Pathol Toxicol Oncol. 2011;30(3):235–40.
21. Finlay JC, Zhu TC, Dimofte A, Stripp D, Malkowicz SB, Whittington R, et al. In vivo determination of the absorption and scattering spectra of the human prostate during photodynamic therapy. Proc SPIE Int Soc Opt Eng. 2014;5315:132–42.
22. Jarvi MT, Patterson MS, Wilson BC. Insights into photodynamic therapy dosimetry: simultaneous singlet oxygen luminescence and photosensitizer photobleaching measurements. Biophys J. 2012;102(3):661–71.

Chapter 8
How to Enter PDT in Clinical Practice?

Cristiane Lassalvia Nascimento, Fábio Parra Sellera, and Martha Simões Ribeiro

Abstract The photodynamic therapy stands out, not only for the few adverse effects to the veterinary patient but also for presenting environmental safety, not inducing microbial resistance and reducing generation of residual drugs on products intended for human consumption. After understanding the full potential of this new therapeutic modality through the previous chapters, the purpose of this chapter is to cover the main aspects to be considered by veterinarians to move PDT to clinical practice. It will be presented how the light source and photosensitizers should be chosen as well as the application modes for both antimicrobial and cancer PDT. Biosafety will also be addressed.

8.1 Introduction

The use of photosensitizers and light to inactivate microorganisms or kill cancer cells is not a novelty; however, this technique did not disseminate in the clinical practice of veterinary medicine. Regarding antimicrobial PDT, possibly it was due to the wide availability of antimicrobial agents developed from the discovery of penicillin [1]. After 2 years of the introduction of penicillin as a therapeutic agent, strains of pathogenic bacteria resistant to penicillin were identified. Thus, a cyclical phenomenon began: as soon as a new antibiotic agent was introduced in the

C.L. Nascimento (✉)
Veterinary Unit of Santos Aquarium, Santos Aquarium, Ponta da Praia, Santos, SP, Brazil
e-mail: cristianelassalvia@hotmail.com

F.P. Sellera
Department of Internal Medicine, School of Veterinary Medicine and Animal Science, University of São Paulo, Butantã, São Paulo, SP, Brazil
e-mail: fsellera@usp.br

M.S. Ribeiro
Center for Lasers and Applications, Nuclear and Energy Research Institute, National Commission for Nuclear Energy, São Paulo, SP, Brazil
e-mail: marthasr@usp.br

F.P. Sellera et al. (eds.), *Photodynamic Therapy in Veterinary Medicine: From Basics to Clinical Practice*, DOI 10.1007/978-3-319-45007-0_8

therapeutic, bacteria resistant to its action mechanism were reported. Probably the notorious phenomenon of microbial resistance, widely revealed in recent years, has made the recent awakening interest in the antimicrobial PDT, even employing photosensitizers known since the Second World War as methylene blue. As yet there is no evidence of microbial resistance to PDT [2], this type of therapy can be extremely valuable in the near future, since the "antibiotic era" may be coming to the end. The application of PDT has been proposed to treat infections superficially localized such as skin infected wounds.

Regarding oncologic PDT, the first substantial animal and human cancer studies were reported by Dougherty and collaborators at the end of 1970s [3]. The initial studies indicated the use of PDT to destroy solid tumors of different sites using porphyrin as the photosensitizer. Today, there is a photosensitizing drug called Photofrin® that is approved by the US Federal Drug Administration (FDA) to be used in cancer treatment. However, with the development of radiation therapy and several chemotherapies, the oncologic PDT was also overlooked. Only in the early 2000s, it returned to spotlight with promising outcome.

Today, the field of PDT itself has evolved considerably so some clinical applications have government approval. This chapter intends to assist those who wish to start employing PDT in their offices prior to the presentation of applications and clinical protocols.

8.2 Choice of Equipment

Today there are various companies concerned with the development of suitable instruments for PDT. The equipment does not have to be only "anatomically appropriate." It must also satisfy requirements that ensure a good combination of quantities of light with the absorption spectrum of the photosensitizer (PS) and also the optical characteristics of the tissue where you want to make the PDT. The type of equipment and clinician's background are directly associated with the success of the PDT. Training is important for the clinician to be able to choose the best wavelength for each illness.

The main requests for PDT light sources are to coincide with the absorption spectrum of the PS, to have adequate power, to be safe in the clinical practice and cost-effective. In terms of lighting, the light sources more recommended are the laser and the LED (light-emitting diode). The most important requirement for an efficient light source in PDT is that a significant value of energy delivered is absorbed by the PS. It need not necessarily be at the peak of the absorption of the PS, but should be efficiently absorbed. It is desirable that the light source has a spectral band that coincides with the absorption band of the PS. If the $A(\lambda)$ represents the absorption spectrum of the PS and $G(\lambda)$ is the spectral density of the source, it is desired that the integral below has the greatest possible value.

$$\int_{0}^{\infty} A(\lambda) G(\lambda) d\lambda$$

If the light needs to penetrate the tissue until reaching the target, we have yet to introduce the function $\xi(\lambda)$, which describes the light penetration through the tissue. Low penetrability indicates that light is absorbed mostly by the tissue cells and does not reach the molecules of the PS. In superficial cases, where there is no need to overcome the biological tissue layers to reach the target, the function $\xi(\lambda)$ is not relevant.

In general, the laser has all energy concentrated in a thin spectral region, being quite easy the determination of the quantity of energy absorbed by the PS. The laser monochromaticity has several advantages in many aspects, but not disqualifies the LEDs in any way. One of the great advantages of laser is its low spatial divergence. Having since its formation a collimated beam, the laser manipulation is easier. If we have to use devices such as optical fibers, laser is undoubtedly the most appropriate source for use.

The LED-based sources have a wider spectral band, being distributed around the central wavelength. Although not collimated, they have spatial distribution of well-known light. Normally, to achieve any collimation, or even to focus the LED light, an effort in well-designed optical components is required.

In situations where it is not necessary the delivery of light in fibers and nor is necessary to transport the light for large distances, the use of LEDs can be of great utility and presents a great advantage when you consider the cost-effectiveness of the device. Suitable light sources for PDT are beginning to emerge on the market and thus are rapidly developing. Both lasers and LEDs are being used in this development.

The light sources on the market differ in wavelength, power, and probe size. Sometimes (e.g., prostate cancer) an optical fiber is necessary to improve the light delivery. For topical use of PDT (e.g., skin infection), blue and green light sources could be employed combined to a PS that absorbs those wavelengths. However, red and near-infrared light sources are more commonly used, mainly for cancer PDT due to higher penetration into the biological tissue associated with the suitable PS (see Chap. 5).

Typically, 100 mW–1 W of useful power is necessary in the red/near-infrared range (600–830 nm) at irradiances ranging from 10 to 200 mW/cm^2 to offer a treatment in tens of seconds (for APDT) or in tens of minutes (for cancer PDT). It is noteworthy that this therapy is nonthermal, i.e., increased local temperature should not be significant.

Lasers present some advantages compared to other light sources. In fact, the coupling into single optical fibers is highly efficient (>90 %) and they are monochromatic, which enable maximum activation of the PS. On the other hand, coupling of LEDs into single optical fibers has shown an efficiency of approximately 50 % (into 600 μm core fiber) and 25 % (into 350 μm core fiber) [4], but LEDs have

Table 8.1 Light sources used in PDT and its advantages and disadvantages

Light sources	Advantages	Disadvantages
Conventional lamps (White – generally emits from UV to IR)	High fluence rate	Filter use
		Temperature rising
		Fiber coupling
		Energy density calculation
LEDs (Blue, green, red, white emission)	Monochromatic	Fiber coupling
	Low thermal effect	
Lasers (Commonly, red and near-IR emission)	Fiber coupling	Less cost-effective than LEDs
	Fluence calculation	Small spot
	Monochromatic	
	Nonthermal effect	

the advantage of presenting low cost compared to lasers and configuring arrays in different irradiation geometries.

Broadband sources (white lamps) can be filtered to match the PS absorbance, but they cannot be efficiently coupled to a single optical fiber, which limits their application for cancer PDT. However, literature reports successful PDT for cutaneous leishmaniasis with the use of 16 % methyl aminolevulinate (PS) and daylight [5].

To calculate the energy or fluence to be used, it is important to know the power of the light source. As seen previously (see Chap. 5), the energy is given by

$$E(J) = P(W) \times t(s)$$

And the fluence can be expressed by

$$F\left(\frac{J}{\text{cm}^2}\right) = \frac{P(W) \times t(s)}{A(\text{cm}^2)}$$

Observe that the power of the light source is given in milliwatts (0.001 W). If you know the energy or fluence to be applied in the treatment, to calculate the exposure time milliwatts must be transformed in watts. On the other hand, the calculation of the area depends on the application, i.e., it could be either the lesion area or the laser beam area. The clinician's decision between scanning or punctual irradiation should be related to the area of the laser beam or the injured area. If the use of an optical fiber is necessary, the area should be calculated using the core fiber diameter.

In surface treatments the light could be used directly from the source with no delivery system, via a lens system, via a fiber bundle/light guide, or via a single fiber optic. Technologies for light delivery include the use of light diffusers specially designed for intracavitary cancer treatment and cylindrically diffusing fibers for interstitial treatments [4].

Table 8.1 presents the main light sources used in PDT.

Table 8.2 Photosensitizers reported in the literature for antimicrobial PDT and their respective absorption bands

Photosensitizer	Absorption band (nm)
Methylene blue	610–660 (red)
Toluidine blue	590–630 (orange/red)
Rose bengal	480–600 (blue to red)
Malachite green	350–450; 550–650 (UV/blue; green to red)
Erythrosine	400–550 (blue to green)
Rhodamine	450–590 (blue to orange)
Violet crystal	420–650 (blue to red)
Porphyrin (e.g., PpIX, HpD)	400–450; 600–650 (blue; red)
Phthalocyanine	660–700 (red/near-infrared)
Curcumin	375–475 (UV/blue)
Chlorin	650–700 (red/near-IR)
Bacteriochlorin	700–800 (near-IR)
Ruthenium-based complexes	<500 nm (UV/blue/green)

PpIX protoporphyrin IX, *HpD* hematoporphyrin derivative, *UV* ultraviolet, *IR* infrared

8.3 Choice of Photosensitizer

For antimicrobial PDT, the literature is rich regarding PDT and PS [6]. The literature reports the use of several types of PS as porphyrins, chlorins, phthalocyanines, curcumin, psoralens, and phenothiazines (see Chap. 3). Table 8.2 shows some PS used for photodynamic inactivation.

For topical infections, the purpose is to kill microorganisms by damaging the cell wall, the structure, and the biochemical composition, which differ for Gram+ and Gram– bacteria and fungi. Thus, the molecular charge of the PS is important to define its diffusion into the microorganism and cationic photosensitizers are desirable. Blue dyes, particularly toluidine blue (TBO) and methylene blue (MB) associated to red light sources show good results in bacteria and fungi inactivation both in vitro and in vivo [6].

The use of MB and TBO is chemically convenient, since the commercial compounds have the same effectiveness as those purified. However, it is noteworthy that those commercials be of high quality and chemical purity about 90 %. Particularly, MB is easy to synthesize and has low toxicity being used in the medical field for a variety of therapeutic purposes such as methemoglobinemia treatment [7], antidote to poisoning by carbon monoxide and cyanide [8], and as a surgical marker in tumor resection [9] in much higher concentrations than those used in PDT, which vary between 50 μM and 1 mM.

The principal classes of clinically applicable PS in cancer PDT are porphyrins, chlorins, and phthalocyanines. The two most commercial porphyrin-based drugs are Photofrin® (or Photogem® in Russia) and Levulan®. Photofrin® (or Photogem®) is a hematoporphyrin derivative (HpD) that presents a large Soret band around 400 nm and some smaller Q bands at longer wavelengths, including a 630 nm peak.

In this wavelength, photons penetrate a few millimeters in skin, so they are indicated for superficial cancers, or tumors that can be touched through endoscopes or optical fibers. Also, it has an extensive residence period in the body, i.e., it is not quickly metabolized, which result in skin photosensitivity for some weeks (4–6 weeks) [4, 6].

Levulan® uses 5-Aminolevulinic acid (ALA), which is not a PS itself, but a precursor for the biosynthesis of endogenous PS protoporphyrin IX (PpIX). PpIX is quickly cleared from the body and, thus, the skin photosensitivity is minimized. Nevertheless, ALA is hydrophilic and its penetration through the skin is limited leading to the development of the M-ALA (methyl aminolevulinate), which is a lipophilic derivative. It is important to highlight that ALA or M-ALA when topically applied provoke significant pain during the light delivery.

It is important to mention Visudyne®, the brand name of verteporfin, which is used to treat mainly age-related macular degeneration and absorbs at $\lambda = 690$ nm. Visudyne is able to occlude mature vessels that may be expressing less or no vascular endothelial growth factor (VEGF).

Looking for PS with longer absorption wavelengths for a high penetration in tissue, researchers have developed chlorin-based PS, considered of second generation. The brand name Foscan® uses verteporfin as the PS, which absorbs red light at $\lambda = 650$ nm. Other marketed PS (Litx™® and Laserphyrin®) use talaporfin, which present maximum absorption wavelength (λ_{max}) at 660 nm. HPPH (2-[1-hexyloxyethyl]-2-devinyl pyropheophorbide-alpha) is being developed as a PS for cancer PDT under the name Photochlor® and its $\lambda_{max} = 665$ nm.

Regarding phthalocyanine-based PS, Photosens® is a mixture of sulfonated aluminum phthalocyanine derivatives that absorbs at $\lambda_{max} = 675$ nm and is being used for a variety of tumors.

The most PS abovementioned are fluorescent and it is possible to monitor PS distribution in the tumor using a different wavelength to excite the molecule. The PS concentration for each treatment will depend on cancer type and is administered to the animal in mg/Kg.

Some commercial PS abovementioned are approved by different regulatory agencies around the world as FDA (US Food and Drug Administration) and European Medicines Agency. Others show potential indication for cancer PDT and are being tested in clinical trials (Table 8.3).

8.4 Application Mode

Another important aspect that must be highlighted is that for antimicrobial PDT the PS application is topical avoiding the photosensitivity to light, which occurs in the case of systemic application of the PS commonly used in cancer PDT (Figs. 8.1 and 8.2).

To perform cancer PDT, a comprehension of PS pharmacokinetics and pharmacodynamics is extremely fundamental to protocol choice. For example, in systemic

Table 8.3 Characteristics of some PS approved for PDT or in clinical trials [10, 11]

Family	Drug	Photosensitizer	~λ_{max} (nm)	Clinical application	Clinical status	Manufacturer	Web site
Porphyrin	Photofrin®	HpD	630	Esophageal cancer, endobronchial cancer, high-grade dysplasia in Barrett's esophagus	Approved in 1995 (FDA)	Axcan Pharma, Inc.	www.photofrin.com
	Photogem®	HpD	630	Esophageal, lung and superficial gastric cancers	Approved in 1999 (Ministry of Health of Russian Federation)	Moscow Research Oncological Institute	www.timtec.net/photogem.html
	Levulan®	PpIX (ALA)	630	Actinic keratosis and superficial basal cell carcinoma	Approved in 1999 (FDA)	DUSA Pharmaceuticals, Inc.	www.dusapharma.com
	Metvixia®	PpIX (m-ALA)	630	Actinic keratosis	Approved in 2004 (FDA)	Galderma S.A.	www.galderma.com
	Visudyne	Verteporfin (benzoporphyrin derivative)	690	Age-related macular degeneration, pathologic myopia or presumed ocular histoplasmosis	Approved in 1999 (FDA)	Novartis Pharmaceuticals	www.visudyne.com

Table 8.3 (continued)

Family	Drug	Photosensitizer	$\sim\lambda_{max}$ (nm)	Clinical application	Clinical status	Manufacturer	Web site
Chlorin	Antrin®	Motexafin lutetium (texaphyrin)	730	Prostate, breast and cervical cancers, arterial disease, age-related macular degeneration	Phases I and II	Pharmacyclics	www.pharmacyclics.com
	Foscan®	Temoporfin	650	Advanced head and neck squamous cell carcinoma	Approved in 2001 (European Medicines Agency)	Biolitec Pharma Ltd.	www.biolitecpharma.com
	Lix™® (Aptocine)	Talaporfin	650	Metastatic colorectal cancer and hepatocellular carcinoma. Studies have been completed or are being planned in solid tumors indications, comprising glioma, nasopharyngeal, head and neck, lung, breast, pancreatic, and prostate cancers	Phases I, II, and III	Light Sciences Oncology, Inc.	www.lsoncology.com
	Photochlor®	HPPH	665	Esophageal and lung cancers, basal cell carcinoma, Barrett's esophagus	Phases I and II	Medkoo Biosciences, Inc.	www.medkoo.com
	Laserphyrin®	Talaporfin	650	Early superficial lung cancer in nonsurgical patients	Approved in 2003 (Japanese government)	Meiji Seika Pharma Co. Ltd.	www.meiji-seika-pharma.co.jp

Bacteriochlorin	TOOKAD®	Padeliporfin	753	Early-stage prostate cancer	Approved in 2016 (COFEPRIS, Mexico)	Steba Biotech Co.	http://stebabiotech.com/index.php/TOOKAD-R-Soluble
Phthalocyanine	Photosen®	Mixture of sulfonated aluminum phthalocyanine derivatives (AlPcS4)	675	Stomach, skin, lips, oral cavity, tongue, and breast cancers	Approved in 2001 (Ministry of Health of Russian Federation)	Niopik	www.niopik.ru

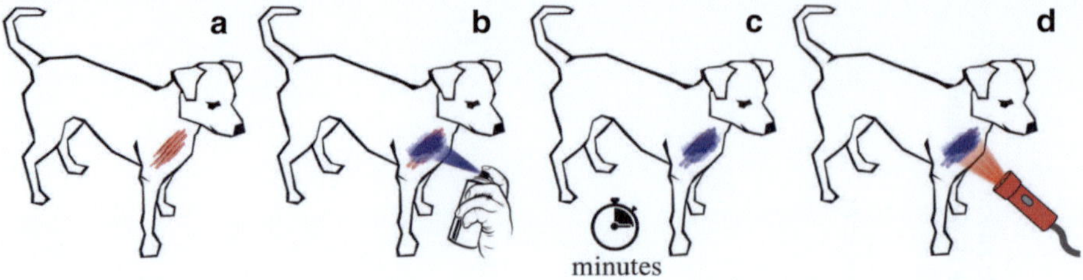

Fig. 8.1 Application mode of PDT for microbial inactivation. (**a**) Infected wound, (**b**) photosensitizer topically administered, (**c**) pre-irradiation time (minutes), (**d**) irradiation procedure

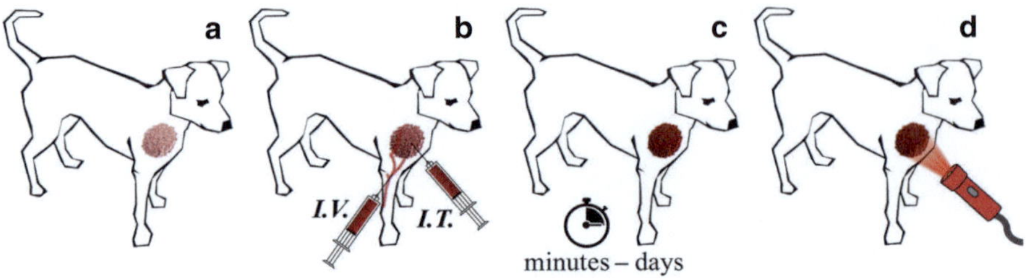

Fig. 8.2 Application mode of PDT for cancer. (**a**) Skin tumor, (**b**) photosensitizer intratumoral (IT) or intravenous (IV) administered, (**c**) pre-irradiation time (minutes–days), (**d**) irradiation procedure

application, before clinician decides what is the most appropriated PS some important aspects as pre-irradiation time, PS elimination time and, primarily, previously to know about PS adverse effects, have to be taken in consideration. Adverse effects in PS systemic or intralesional administration may include cutaneous photosensitivity posttreatment, local pain, edema, lethargy, photophobia, hepatic necrosis, etc. [12]. PS topical administration to treat infections promotes few adverse effects, mainly because the most commonly used (MB and TBO) are considered safe for the host cells and ALA has a short half-life (24–48 h). Furthermore, with a proper cleaning, to remove the PS is relatively easy and simple.

8.5 Biosecurity

Biosecurity is the set of actions for preventing, minimizing, or eliminating the risks inherent to research, technological development, and provision of services to the health of humans, animals, and the preservation of the environment and the quality of results. Biosafety concepts should also be respected both in clinical application as laboratory PDT.

As the PDT requires the use of radiation sources, safety and protection that apply to the use of lasers and also other intense radiation sources must be respected. The

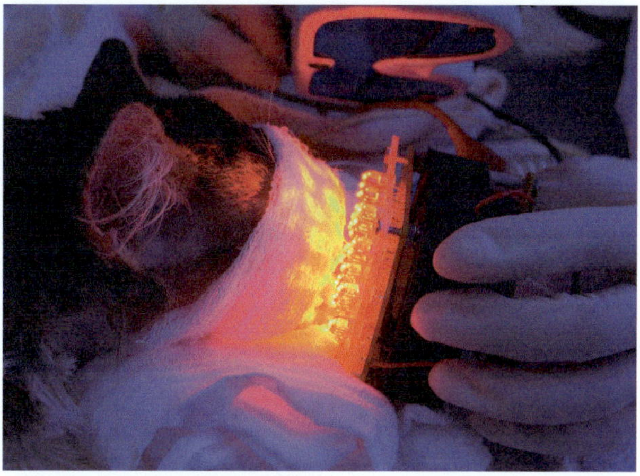

Fig. 8.3 Bandage gauze used to protect patient eyes during irradiation procedure (Image kindly provided by Claudia Rodriguez Emilio)

use of goggles is obligatory in any laser practice. The operating parameters of your laser will define what forms of protection you should have in place.

Some lasers emit in the range of invisible light. In this case, commonly a low-power red light is used as a "guide." Low-power visible lasers can be harmful when pointed to the eyes since the beam can be focused in retina. Dark glasses that do not allow the passage of visible radiation should be used. The clinicians who will perform the irradiation as well as the support staff that is next to the application site should use glasses that stop the wavelength being used, but allow the visualization of the operative field. It is also recommended good sense on protection for animals. For example, physical barriers (e.g., gauze bandage or similar) could be used when PDT will be performed close to eyes (e.g., carcinoma or skin infection on face region) (Fig. 8.3). Chapter 10 will also describe alternative shields for the eyes.

Another point to be discussed is the use of anesthesia procedures. Obviously, to keep animals motionless will always be better. Moreover, cancer PDT could promote pain reactions related to PS administration (e.g., intralesional) or during PDT procedure [13]. Thus, we strongly encourage the presence of an anesthesiologist in this kind of procedure. On the other hand, in our experienced antimicrobial PDT to treat skin infections, we did not observe pain reactions in our patients. So, we believe that physical restraint (when it is possible) or conditioning (especially when treatment needs to be performed repeatedly) is required.

The veterinarian biosafety is basically supported by individual protection (professional and animal) and disposal of waste. In the specific case of PDT, individual protection requires specific protective eyewear beyond conventional safety equipment (gloves, mask, mob caps, etc.). It is important the cleaning and disinfection of the equipment and cables associated with the light source. The use of plastic protective film on the tip of the light source should be encouraged, but it must be done properly so that the barrier does not interfere significantly with the radiation emitted by the equipment.

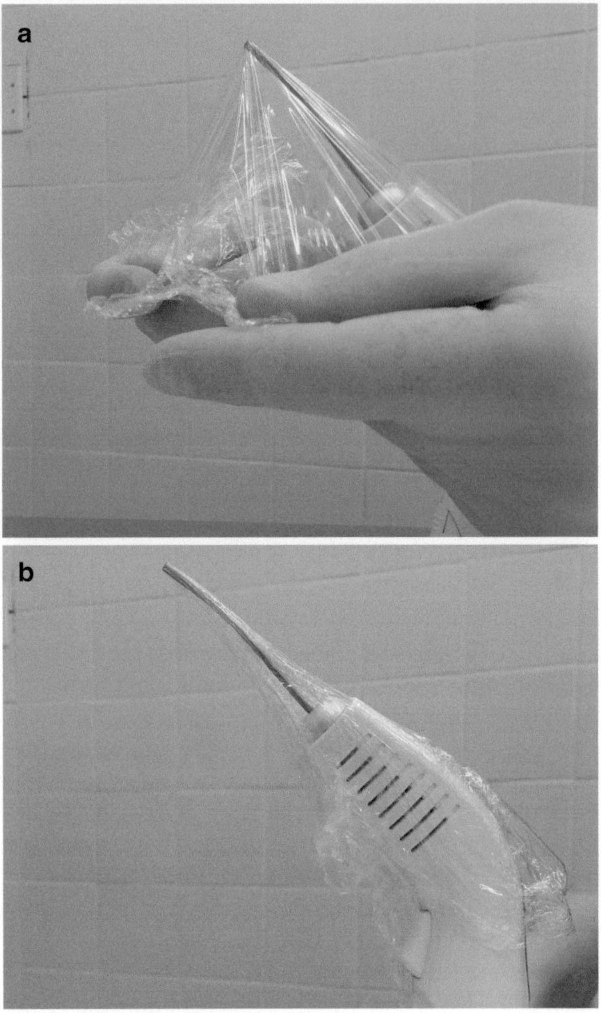

Fig. 8.4 Protection of laser equipment. (**a**) A stretch film must be used to involve the laser tip; (**b**) plastic film covering all parts of the laser device to avoid contamination

In addition, the cleaning of the equipment should be a concern for users. Some laser devices have autoclavable tips. The remainder should be decontaminated by friction with 70 % ethanol and protected with plastic film before use (Fig. 8.4).

The optical fibers used and developed for PDT are heat sensitive, so they must be sterilized chemically. According to the manufacturer, these fibers should not be reused but discarded after use. It is important to note that the fiber in contact with the photosensitizer leads to formation of a thin barely colored layer, which persists after rinsing with water and cleaning with 70 % ethanol-moistened gauze.

The photosensitizing solution also requires special attention because the light absorption can generate reactions inside the bottle. Thus, the photosensitizer storing must be performed in dark preferably refrigerated. To ensure no contamination of the PS solution, its removal from the bottle with a sterile syringe is recommended and correct closure and storage of the bottle should be proceeded. In the case of transparent bottles, use an aluminum foil or a black tape to protect the PS from light (see Chap. 10).

References

1. Hamblin MR, Hasan T. Photodynamic therapy: a new antimicrobial approach to infectious disease? Photochem Photobiol Sci. 2004;3(5):436–50.
2. Jori G, Fabris C, Soncin M, Ferro S, Coppellotti O, Dei D, et al. Photodynamic therapy in the treatment of microbial infections: basic principles and perspective applications. Lasers Surg Med. 2006;38(5):468–81.
3. Dougherty TJ, Kaufman JE, Goldfarb A, Weishaupt KR, Boyle D, Mittleman A. Photoradiation therapy for the treatment of malignant tumors. Cancer Res. 1978;38(8):2628–35.
4. Wilson BC, Patterson MS. The physics, biophysics and technology of photodynamic therapy. Phys Med Biol. 2008;53(9):R61–109.
5. Enk CD, Nasereddin A, Alper R, Dan-Goor M, Jaffe CL, Wulf HC. Cutaneous leishmaniasis responds to daylight-activated photodynamic therapy: proof of concept for a novel self-administered therapeutic modality. Br J Dermatol. 2015;172(5):1364–70.
6. Yin R, Hamblin MR. Antimicrobial photosensitizers: drug discovery under the spotlight. Curr Med Chem. 2015;22(18):2159–85.
7. Savino F, Maccario S, Guidi C, Castagno E, Farinasso D, Cresi F, et al. Methemoglobinemia caused by the ingestion of courgette soup given in order to resolve constipation in two formula-fed infants. Ann Nutr Metab. 2006;50(4):368–71.
8. Betten DP, Vohra RB, Cook MD, Matteucci MJ, Clark RF. Antidote use in the critically ill poisoned patient. J Intensive Care Med. 2006;21(5):255–77.
9. Golshan M, Nakhlis F. Can methylene blue only be used in sentinel lymph node biopsy for breast cancer? Breast J. 2006;12(5):428–30.
10. Ormond A, Freeman H. Dye sensitizers for photodynamic therapy. Materials (Basel). 2013;6(3):817–40.
11. Sekkat N, van den Bergh H, Nyokong T, Lange N. Like a bolt from the blue: phthalocyanines in biomedical optics. Molecules. 2012;17(1):98–144.
12. Peaston AE, Leach MW, Higgins RJ. Photodynamic therapy for nasal and aural squamous cell carcinoma in cats. J Am Vet Med Assoc. 1993;202(8):1261–5.
13. Buchholz J, Walt H. Veterinary photodynamic therapy: a review. Photodiagnosis Photodyn Ther. 2013;10(4):342–7.

Chapter 9
Basic Studies in Cancer PDT

Julia Buchholz

Abstract The first reports on photodynamic therapy (PDT) date back to the beginning of the last century, when researchers observed that a combination of light and certain chemicals could induce cell death. Significant efforts have been invested in the development of sensitizers to optimize the treatment since the photosensitizer is considered a critical element in PDT and should therefore at least meet some of the following criteria that are clinically relevant. Lots of sensitizers selectively localize within a tumor due to physiological differences in the tumor and healthy tissue; rather long-wavelength absorbing photosensitizers are preferentially used in oncologic PDT. The disadvantage of topical PDT in oncology is a limited penetration of the photosensitizer until few millimeters. Therefore mainly superficial tumors needing treatment up to 1 cm depth are amenable to PDT, with the exception being interstitial PDT or a combination of PDT with prior debulking surgery. In addition to the photodynamic therapeutic effect that occurs after light administration, photosensitizers can also be used diagnostically since many of these substances induce fluorescence. It is likely that PDT will continue to be used as both a solitary treatment modality and in combination with other strategies, such as hyperthermia and photodynamic therapy-generated vaccines.

9.1 Introduction

The first reports on photodynamic therapy (PDT) date back to the beginning of the last century, when researchers observed that a combination of light and certain chemicals could induce cell death. Nowadays, these chemicals are known as photosensitizers. Light energy transferred to the photosensitizer will lead to a chemical reaction, and in the presence of molecular oxygen, singlet oxygen ($^1O_2$) or superoxide (O_2^-) will be produced and will induce cell damage. Cell destruction after PDT can occur in several ways: (1) direct tumor cell destruction (via apoptosis or

J. Buchholz
Animal Oncology and Imaging Center, Huenenberg, Switzerland
e-mail: buchholz@aoicenter.ch

F.P. Sellera et al. (eds.), *Photodynamic Therapy in Veterinary Medicine: From Basics to Clinical Practice*, DOI 10.1007/978-3-319-45007-0_9

necrosis), (2) damage to the tumor vasculature, and (3) modulation of the immune system [1].

Significant efforts have been invested in the development of sensitizers to optimize the treatment since the photosensitizer is considered a critical element in PDT and should therefore at least meet some of the following criteria that are clinically relevant.

A high selectivity toward neoplastic tissue is one of the important features characterizing a photosensitizer used in oncology. Lots of sensitizers selectively localize within a tumor due to physiological differences in the tumor and healthy tissue [2]. Interaction of photosensitizers with tumors can happen via low-density lipoprotein (LDL) receptors, which are elevated in cancer cells and therefore endocytosis of LDL-photosensitizer complex is preferred by malignant cells [3]. A high fraction of tumor-associated macrophages is found in neoplastic cells also leading to high photosensitizer levels in these areas [4]. Selective uptake by tumor cells is possibly due to lower intracellular pH, leaky microvasculature, poor lymphatic drainage by tumors, and large amounts of collagen [5]. While water solubility is important for bioavailability of the sensitizer, lipophilicity is important for diffusion through lipid barriers and localization in endocellular sites [6]. Liposomal formulations can also increase drug selectivity toward a higher tumor-to-normal tissue ratio [7]. Achieving a maximum tumor-to-normal cell concentration ratio can take few hours/days, depending on the photosensitizer and the tumor type.

Rather long-wavelength absorbing photosensitizers are preferentially used in oncologic PDT (therapeutic window: 600–800 nm) because at low wavelengths absorption of light by tissue is high. On the other hand, negative effects, such as photosensitizer photobleaching, are also possible at wavelengths far into the red or near infrared regions [8].

Possible administration through various routes can be an advantage: a sensitizer can be administered intravenously, intraperitoneally, or topically. The disadvantage of topical PDT in oncology is a limited penetration of the photosensitizer until few millimeters. In general, the poor penetration into tissue (due to photosensitizer and/or light) is one of the main limitations for treating tumors with PDT. Therefore, mainly superficial tumors needing treatment up to 1 cm depth are amenable to PDT, with the exception being interstitial PDT or a combination of PDT with prior debulking surgery. With systemic PDT, the penetration depth usually is increased compared to topical application and is mainly defined by the wavelength (see above). Most important is the optimization of PDT protocols such as drug and light dosage as well as the time between the administration of the drug and the light exposure (drug-light interval). A short drug-light interval might be convenient to enable performing PDT in 1 day. Also, short-generalized skin photosensitivity is advantageous in particular in veterinary medicine since it might be a limitation to keep the patients inside after treatment for an extended period of time.

In addition to the photodynamic therapeutic effect that occurs after light administration, photosensitizers can also be used diagnostically since many of these substances induce fluorescence. Following photosensitizer localization, fluorescence from the sensitizer can be used for tumor detection. An example is the photodynamic diagnosis of breast tumors or diseased axillary sentinel lymph nodes after oral application of 5-aminolevulinic acid (5-ALA) [9], a frequently for diagnostics and therapy used prodrug [10]. In veterinary medicine, preliminary results of sentinel lymph node mapping of invasive urinary bladder cancer are available [11].

Many photosensitizers were introduced in the 1980s and 1990s, and in 1995, the Federal Food and Drug Administration (FDA) approved PDT as an alternative form of therapy against cancer. Suitable sensitizers for PDT are mainly porphyrinoid compounds (porphyrins, chlorins, bacteriochlorins, and phthalocyanines). They absorb light in the visible region making them colored compounds or dyes. Non-porphyrins are, for example, anthraquinones, phenothiazines, xanthenes, cyanines, and curcuminoids [12].

Photofrin® (porfimer sodium) is a hematoporphyrin derivative (HpD) and it was the first FDA-approved PDT sensitizer. These photosensitizers are known as first-generation photosensitizers. They result in long-term skin phototoxicity (lasting 6–10 weeks), meaning sunlight and strong artificial light exposure must be avoided during this period. Although Photofrin® has its weaknesses, it still is a potent and safe photosensitizer, and it was approved in 1993 by Canada for treatment of bladder cancer and by the US FDA for treating esophageal cancer in 1995, lung cancer in 1998, and Barrett's esophagus in 2003. Photofrin® treatment extends to head, neck, abdominal, thoracic, brain, intestinal, skin, breast, and cervical cancer. Other types of hematoporphyrin derivatives are Photogem® and Photosan-3®. Photogem® has been approved for use in clinical applications in Russia and Brazil. Photosan-3® has been approved for clinical use in the EU [12].

Second-generation photosensitizers were explored among others because of the long-lasting skin phototoxicity and a low absorption in the red region of the visible spectrum of the first-generation photosensitizers. Endogenous protoporphyrin IX (PpIX) induced by exogenous 5-aminolevulinic acid (ALA or Levulan Kerastick®) was US FDA approved for non-oncological PDT treatment of actinic keratosis in 1999 [13]. Via the heme pathway, PpIX, which acts as a photosensitizer, is produced enzymatically from the prodrug ALA. Due to reduced ferrochelatase in neoplastic tissue, excess ALA leads to accumulation of PpIX in the mitochondria. The PpIX absorption maximum is about 630–635 nm. The methyl ester of ALA, methyl aminolevulinate (MAL, Metvix®, or Metvixia®), was approved by the US FDA in 2004 for treatment of actinic keratosis [12]. Under the trade name Metvixia®, MAL is also used as a topical treatment and has an advantage over Levulan® due to the nature of the irradiation source. BLU-U® light was approved for use with Levulan® as the most efficient source emitting at 400 nm, while Aktilite® was approved for Metvixia® which emits at 630 nm and provides deeper tissue penetration. Hexaminolevulinate, the *n-hexyl* ester of ALA (HAL, Hexvix®, Cysview®), was approved in 2010 by the US FDA in the diagnosis of bladder cancer [12]. HAL is converted to PpIX 50–100 times more efficiently than ALA. Clinical trials are ongoing for treatment of cervical intraepithelial neoplasia [14] and genital erosive lichen planus [15].

Several photosensitizers evaluated for PDT efficacy are from the chlorin family and include benzoporphyrin derivative monoacid ring A (BPD-MA, verteporfin, Visudyne®), *meta-tetra(hydroxyphenyl)chlorin* (*m*THPC, Foscan®), tin ethyl etiopurpurin (SnET2, rostaporfin, Purlytin™), and *N-aspartyl* chlorin e6 (NPe6, talaporfin, LS11) which is derived from chlorophyll a. When compared to porphyrins, the structure of chlorins differs by two extra hydrogens in one pyrrole ring. This structural change leads to a red shift in the absorption band to higher wavelengths (640–700 nm). BPD-MA is activated by light at 689 nm and has a lower time interval of skin phototoxicity than Photofrin®, due to rapid plasma and tissue pharmacokinetics, which enables faster excretion of the drug from the body [16]. In 1999, US FDA approved the use of BPD-MA as Visudyne® for age-related macular degeneration in ophthalmology [18]. Additionally, a 24-month study of verteporfin treatment showed improvement in patients with nonmelanoma skin cancer [17].

PDT treatment of neck and scalp cancer with *mTHPC* was approved in Europe, and the drug was used successfully for treating breast, prostate, and pancreatic cancers [18, 19]. Light activation at 652 nm is very effective and only small doses of *mTHPC* are required during treatment. A weakness of *mTHPC* is high skin photosensitivity in some patients. SnET2, under the trademark Purlytin™, has been evaluated in Phase I/II trials for the treatment of metastatic breast adenocarcinoma, basal cell carcinoma, and Kaposi's sarcoma [20]. This drug has also finished Phase III trials for the treatment of age-related macular degeneration but has not yet been approved by the FDA, due to a requirement of further efficacy and safety assessments [21]. Purlytin™ is activated at 664 nm and has deeper tissue penetration than Photofrin®. The drawback of the drug is a possibility of dark toxicity and skin photosensitivity.

Several other promising photosensitizers are currently under clinical trials. These include HPPH (2-[1-hexyloxyethyl]-2-devinyl pyropheophorbide-a, Photochlor), motexafin lutetium (MLu, lutetium(III) texaphyrin, Lu-Tex), NPe6 (mono-L-aspartyl chlorin e6, taporfin sodium, talaporfin, LS11), and SnET2 (tin ethyl etiopurpurin, Sn etiopurpurin, rostaporfin, Photrex) [22].

Another important requirement for PDT is light. The first light sources used in PDT were noncoherent light sources. These are safer, easy to use, and less expensive. Since they produce spectra of wavelengths, they can accommodate various photosensitizers. Using optical filters, light of selective wavelength(s) can be produced. Some disadvantages of conventional lamps used to be significant thermal effect, low light intensity, and difficulty in controlling light dose. Today, most of these drawbacks can be overcome by a careful engineering design [23]. An example is an illumination system for ALA-PDT of actinic keratosis (AK) of the face and scalp (BLU-U light illuminator (DUSA Pharmaceuticals, Inc.)), a simple timer-controlled switch on-off unit. The LumaCare™ lamp (MBG Technologies), a compact portable fiber-optic delivery system, provides interchangeable fiber-optic probes containing a series of lenses and optical filters. It can generate light of specific bandwidth between 350 and 800 nm in a variable power for a broad range of photosensitizers [23]. Light emitting diode (LED) is another emerging PDT light applicator used for PDT procedures. LED can generate high-energy light of desired wavelengths and can be assembled in a range of geometries and sizes. For intraoperative PDT of brain tumors, the LED probe may be arranged in a cylinder tip to fit into a balloon catheter [24], whereas, for minimally invasive interstitial PDT, the small and flexible light delivery LED catheter can be implanted into the tumor percutaneously [25]. Large LED arrays may be more suitable for flat surface illumination of wide-area superficial lesion [26].

Lasers are still the most commonly used PDT light sources producing high-energy monochromatic light of a specific wavelength with a narrow bandwidth for a specific photosensitizer. The laser light can be focused, passed down an optical fiber, and directly delivered to the target site through a specially designed illuminator tip, for example, a microlens or a cylindrical or spherical diffuser. Dose planning is more complex with intracavitary or interstitial illumination, mainly due to backscattering. Some effort was put into the development of flexible textile light diffusers with the aim of a homogeneous illumination in body cavities or the oral cavity, but dosimetry is quite challenging [27, 28]. Argon dye, potassium-titanyl-phosphate (KTP) dye, metal vapor lasers, and most recently diode lasers have been used for clinical PDT around the world. The KTP-dye modular combination system (Laserscope PDT Dye Module) was the most widely used PDT laser prior to the approval of the portable, light-weight, and less expensive diode lasers (e.g., DIOMED 630 PDT; Diomed Inc.). One preferred advantage of the diode laser is that it can be engineered into a multi-channel unit to meet a highly sophisticated PDT procedure, which may require multi-channel diode lasers and each independent light output channel to simultaneously provide the light sources of variable power (e.g., Ceralas PDT 762 nm; CeramOptec GmbH of Biolitec AG) [23].

It is likely that PDT will continue to be used as both a solitary treatment modality and in combination with other strategies, such as hyperthermia [29] and photodynamic therapy-generated vaccines [30]. Concerning the former, it can be said that the simultaneous application of hyperthermia and PDT seems to react synergistically [31], but the dosimetry of hyperthermia treatment is very challenging. In a recent study, photodynamic hyperthermal chemotherapy was used to treat 16 patients (10 dogs and 6 cats) with malignant soft tissue sarcoma. They conclude that this treatment combination could decrease the risk of STS recurrence [32].

Concerning the use of PDT-generated vaccines, it has been shown that tumor cells treated in vitro by PDT can be used for generating potent cancer vaccines [33, 34]. Injection of such whole-cell PDT vaccine into mice-bearing tumors of the same origin as used for the vaccine generation was shown to produce significant antitumor effect even with models of poorly immunogenic carcinoma [34]. The tumor specificity could be demonstrated by an ineffectiveness against mismatched tumors [33, 34]. Cytotoxic T lymphocytes play a key role in the destruction of tumors mediated by the PDT vaccine. In lesions regressing after PDT vaccine treatment, high numbers of degranulating CD8+ cells were found after PDT, versus much lower numbers in the tumors within the same treatment group that exhibited a poor early response (progressing after vaccination). Elevated numbers of degranulating CD8+ cells were also found in tumors treated by regular PDT [35]. While PDT vaccines injected at a distal site were still effective, their impact was inferior to perilesional treatment. The proximity of vaccination site to the treated lesion is relevant for the therapy outcome even though the PDT vaccine-induced antitumor immune response is of systemic nature (and cured mice resist tumor rechallenge). Antigen presentation to T cells takes place in the lymph nodes nearest to the vaccine injection site, and in case of peritumoral vaccine injection, T-cell activation will be centered in tumor-draining lymph nodes, which is a location favorable for T-cell trafficking into the tumor. In addition, it could be shown that the potency of PDT vaccine is increased if vaccine cells remain in culture after PDT treatment for an additional time interval to allow the expression of PDT-induced molecular/biological changes within these cells [35]. Such changes of possible relevance include the expression of genes in PDT-treated cells whose products are important immune response mediators (e.g., heat shock proteins), which appear to play a key role in the action of PDT vaccines [34]. PDT vaccines are more effective in eradicating smaller tumors, but Korbelik et al. emphasize that this does not mean that they cannot be employed for therapy of larger lesions. Decreasing the tumor burden by other treatment strategies or combination with adjuvant immunomodulatory treatment could be helpful for future studies and eventually use in a clinical setting. Veterinary patients might be a very useful model in this setting resulting in a win-win situation for our patients and human patients. An important practical result of this study is that tumor tissue can be directly used for the production of PDT vaccine without the need for generating first cultures of single cancer cells. Surgically removed tumor tissue from the patient can be directly used for preparing the PDT vaccine material tailored for the individual patient which is acting against tumor antigens existing in that specific tumor [30].

From 1975 on, the use of PDT in animals with cancer has been published [36]. Until now many PDT publications and some reviews [37, 38] are available about PDT in veterinary medicine. So far, different photosensitizers have been used therapeutically in veterinary medicine to treat different tumors in different species.

Pharmacokinetic studies of some of these substances have been performed: HPPH (Photochlor; 2-(1-hexyloxyethyl)-2-devinyl pyropheophorbide-*a*) [39–41], $AlPcS_4$ (aluminum phthalocyanine tetrasulfonate) [42], *m*THPC (*meta*-tetrahydroxyphenylchlorin) [7, 43, 44], $ZnPcS_4$ (zinc phthalocyanine tetrasulfonate) [45]. Some of the studies cannot be clearly defined as being a basic or a clinical study.

9.1.1 Basic Studies in Veterinary Medicine

In 1996, the tissue biodistribution and photodynamic effects of *m*THPC on normal canine prostate was studied in vivo. Highest *m*THPC concentration was seen 24–72 h after IV administration. Interstitial PDT led to lesions (swelling, inflammatory response, and extensive glandular destruction) up to 4 cm in diameter (using four fiber sites). Interestingly, there was no disruption of the main stroma, and size and shape of the gland remained unchanged, whereas persistent glandular atrophy could be observed at 90 days. No animal became incontinent, but temporary urinary retention was caused by urethral damage until 7 days posttreatment. In some animals, small rectal lesions developed, which healed completely subsequently without formation of fistulae. These results demonstrate the potential use of PDT for prostatic cancer limited to the gland since glandular tissue can be selectively destroyed. The use for benign prostatic hyperplasia does not seem to be promising because of maintenance of the structural integrity of the prostate [46].

In 1999, the same group studied, in addition to *m*THPC, the biological responses of canine prostate to a phthalocyanine photosensitizer (AlS2Pc) as a preparatory step for clinical trials. Having the animals protected from light for 2 weeks following photosensitizer injection, skin photosensitivity could be neglected. *m*THPC induced more prostate lesions than AlS2Pc. Probably due to severe urethral irritation and acute swelling of the prostate, dogs were in physical distress and regeneration of urethral epithelium was not complete until 3–4 weeks after PDT. Also in this study, they could show that the glandular architecture remained well preserved because the interlobular collagens were less affected than the cellular components of the gland. Importantly, both photosensitizers could preserve the dogs from developing periprostatic nerve damage or rectal lesions. The authors conclude that the long-term therapeutic effectiveness for prostate cancer needs further investigation [47].

In 1997, a preclinical study about sequential whole bladder PDT was conducted in six female dogs using 1.5 mg/kg Photofrin® and 15 J/cm^2 of light (low-dose Photofrin® PDT). Two dogs received one PDT, four dogs received two PDTs, and three dogs received three PDTs. A single PDT resulted in average bladder capacity losses of 11 % (0–33) and 0 % at weeks 1 and 12, respectively. Two treatments induced average bladder capacity losses of 36 % (0–57) and 17 % [2–24] at weeks 1 and 12, respectively. Three sequential treatments resulted in average bladder capacity losses of 22 % (0–42) and 0 % at weeks 1 and 12, respectively. Except after the second sequential PDT, full recovery in bladder capacity occurred in all cases. The authors therefore conclude that this protocol of sequential whole bladder low-dose Photofrin® PDT is safe to be used in clinical investigation [48].

A preclinical toxicity study from 2003 investigated the photosensitizer HPPH (0.3 mg/kg) or Photofrin® (2.0 mg/kg) intravenously 48 h prior to thoracotomy and light delivery for potential treatment of sarcoma pulmonary metastases. Fourteen beagles received one of the photosensitizers IV 48 h prior to thoracotomy and light delivery (interstitially or pleurally). Significant toxicities occurred after pleural PDT using 15 and 30 J/cm^2 including pyrexia and respiratory distress. Using single beam pleural surface treatment, all light doses (5–40 J/cm^2) were tolerated for both photosensitizers with depths of treatment effect up to 10 mm (with HPPH). Therefore, after adjusting light dosimetry, histologic and clinical changes associated with interstitial and pleural PDT were deemed acceptable, and the authors conclude that PDT might be a promising adjuvant treatment for sarcoma pulmonary metastases [49].

Gloi and Beck studied photodynamic threshold doses of porfimer sodium, AlClPc, and SnET2 to overcome the problem of variability between different photosensitizers and species [50]. In 12 dogs with spontaneous tumors, the tissue optical properties of each compound were determined by diffuse reflectance. Threshold values were highest for AlClPc. Therefore, more photon absorption is needed for tumor necrosis with AlClPc than for either SnET2 or porfimer sodium. In a second study, they found a marked variability among species in the distribution of porfimer sodium between highly proliferating tissues demonstrating the necessity for careful attention in the design of human and veterinary application of PDT [50, 51]. The same might be true for other photosensitizers.

A preclinical study using interstitial motexafin lutetium (MLu)-mediated PDT for recurrent prostate cancer was conducted in 16 dogs and in 16 humans in a Phase I study. The treatment was safe, but there was significant variability in dose distribution and subsequent tissue necrosis. In order to deliver a consistent overall PDT dose, an integrated system being able to detect and compensate for variation in dose distribution in real time would be necessary to guarantee therapeutic reproducibility [52].

Motexafin lutetium (MLu) was used in another study for PDT for rectal cancer. This preclinical study included ten dogs as a model for treatment of rectal cancer. Eight dogs underwent proctectomy and low rectal end-to-end stapled anastomosis. Six dogs received MLu at 2 mg/kg preoperatively and underwent subsequent pelvic illumination of the transected distal rectum (730 nm light with light dose from 0 to 5–10 J/cm^2 3 h after drug delivery). Two dogs received light but no drug and underwent proctectomy and low rectal stapled anastomosis. Two dogs underwent midline laparotomy and pelvic illumination. No significant tissue toxicity could be observed, but mild enteritis was found histologically in all dogs. Low rectal stapled anastomosis performed with MLu pelvic PDT is safe in a dog model. The authors conclude that 730 nm light administered in pelvic tissue can treat residual disease of less than 5 mm adequately with MLu-mediated PDT and merits further investigation as an adjuvant therapy in this setting [53].

In a pharmacokinetic study, benzoporphyrin derivate monoacid ring A (BPD-MA, verteporfin) was administered intravenously in nine dogs with naturally occurring tumors. The $t_{1/2}$ was longer than in humans but similar to the $t_{1/2}$ found in rats. No clinical abnormalities, except for transient increase in serum alkaline phosphatase activity, were observed. They conclude that the pharmacokinetics of BPD-MA in tumor-bearing dogs would be helpful in determining the protocol of a short drug-light interval PDT mainly targeting the tumor vasculature [54].

In a Phase I clinical trial, the safety of zinc phthalocyanine tetrasulfonate (ZnPcS(4)) was evaluated in pet dogs with naturally occurring tumors. Drug doses of $\leq$4 mg/kg induced no toxicity and resulted in partial to complete tumor responses in 10/12 dogs 4 weeks after PDT. Tumor remission was observed with ZnPcS(4) doses as low as 0.25 mg/kg. The authors therefore conclude that the identification of the maximum tolerated dose through traditional Phase I clinical trials may be unnecessary for evaluating novel PDT protocols [45].

Interstitial PDT using an intra-arterial photosensitizer (benzoporphyrin derivate 1, 3-diene C, D-diethylene glycol ester A ring (QLT0074)) and computerized pulsed light was administered to the prostate of 11 dogs. Two mg of the photosensitizer was given per prostaticovesical artery bilaterally. Immediately following infusion, the prostate was surgically exposed and seven optical fibers with 1.5 cm cylindrical diffusers after loading sheaths were inserted into the prostate through a template. Light was delivered sequentially to the optic fibers via a computer-driven switch system. One dog was sacrificed 6 days after PDT and comprehensive destruction of the prostate was seen. The other ten dogs were monitored for clinical tolerance and urinary function and sacrificed 3–11 months after PDT. Except for urinary retention and mild hematuria, no other complications could be observed. The authors concluded that the treatment was feasible and safe resulting in destruction of the prostate [55].

In two in vitro studies, the effect of PDT with hematoporphyrin monomethyl ether (HMME) on canine breast cancer cells (CHMm cell line) was examined. The first study could show that apoptosis plays a major role in PDT-HMME-induced reduction of the viability of CHMm cells. The second study demonstrated damages of the mitochondrial structure and mitochondrial dysfunction [56, 57].

In another study, PDT with aluminum-chloride-phthalocyanine encapsulated in liposomes on canine breast cancer cells was determined. They could show decreased cell viability with morphologic alterations post PDT. The tumor cell destruction was predominantly mediated via a necrotic process [58].

A recent safety study using sinoporphyrin sodium (DVDMS) was conducted in beagle dogs. The dogs were randomly allocated to six groups and a DVDMS preparation was given IV at dose levels from 0 to 9 mg/kg. A group receiving 1 mg and 9 mg, respectively, was illuminated 24 h later with a 630 nm laser for 10 min, once every 7 days for 5 weeks. Skin swelling and ulceration were seen in dogs that received PDT indicating that DVDMS-PDT induced a phototoxic effect. Pigmentation of DVDMS (or its metabolite) was observed to deposit in the liver, spleen, local lymph nodes, and bone marrow of dogs in the mid- and high-dose groups, as well as in the high-dose PDT group. The targets therefore are presumed to be the liver and immune organs. The no-observed-adverse-effect level (NOAEL) was considered to be 1 mg/kg and the treatment seemed to be safe and promising [59].

The prodrug 5-aminolevulinic acid (5-ALA) has been studied in veterinary medicine as well. In cats, 5-ALA is used mainly topically due to observed hepatotoxicity after intravenous administration [60].

In vitro studies have shown that ALA destroys canine transitional cell carcinoma cells [61]. Even though 70 % of dogs vomited after oral administration of ALA, this did not have a negative impact on pharmacokinetics: A ten times higher accumulation of the active metabolite could be found in the mucosa, compared to muscularis and serosa, which should enable a selective treatment. Five dogs with transitional cell carcinoma of the urinary bladder [62] and one dog with a prostatic carcinoma [61] showed transient improvement of clinical symptoms.

Two studies using the palladium bacteriopherophorbide photosensitizer TOOKAD® and laser light to treat canine prostates resulted in an effective ablation of prostatic tissue and might even be used for patients in which radiation therapy for prostatic carcinoma failed. In the first study, the prostate of the canine patients were exposed to ionizing radiation (54 Gy) 5–6 months prior to interstitial TOOKAD®-mediated PDT. Light irradiation (763 nm, 50–200 J/cm at 150 mW/cm from a 1 cm cylindrical diffusing fiber) was delivered during IV infusion of TOOKAD® at 2 mg/kg over 10 min. One week after PDT, the prostates were histologically examined and hemorrhagic necrosis was found (clearly distinguishable from the radiotherapy-induced preexisting fibrosis). No noticeable damage to the urethra, bladder, or adjacent colon could be observed. TOOKAD®-PDT might be an option for patients with recurrent prostate cancer after radiotherapy. In the second study, the authors describe TOOKAD®-PDT as being very effective in ablating prostatic tissue through its vascular effects [22, 63]. Another study presents preliminary results of intraoperative PDT with 5-ALA in dogs with prostatic carcinoma. The median survival time was only 41 days – one of the reasons for the poor outcome might have been the insufficient light penetration using a halogen broad band lamp for illumination instead of a laser [64].

This overview illustrates the complexity of the interaction of photosensitizer, tumor type, species, light source, wavelength, light intensity, drug-light interval, etc., to render PDT a successful yet well-tolerated and feasible cancer treatment.

References

1. Moor ACE, Ortel B, Hasan T. Mechanisms of photodynamic therapy. In: Photodynamic therapy. Cambridge: Royal Society of Chemistry; 2003. p. 19–58.
2. Moan J, Peng Q. An outline of the history of PDT. In: Patrice T, editor. Photodynamic therapy. Cambridge: Royal Society of Chemistry; 2003. p. 1–18.
3. Jori G, Reddi E. The role of lipoproteins in the delivery of tumour-targeting photosensitizers. Int J Biochem. 1993;25(10):1369–75.
4. Korbelik M. Low density lipoprotein receptor pathway in the delivery of Photofrin: how much is it relevant for selective accumulation of the photosensitizer in tumors? J Photochem Photobiol B. 1992;12(1):107–9.
5. Henderson BW, Dougherty TJ. How does photodynamic therapy work? Photochem Photobiol. 1992;55(1):145–57.
6. Jori G. Tumour photosensitizers: approaches to enhance the selectivity and efficiency of photodynamic therapy. J Photochem Photobiol B Biol. 1996;36(2):87–93.
7. Buchholz J, Kaser-Hotz B, Khan T, Rohrer Bley C, Melzer K, Schwendener RA, et al. Optimizing photodynamic therapy: in vivo pharmacokinetics of liposomal meta-(tetrahydroxyphenyl)chlorin in feline squamous cell carcinoma. Clin Cancer Res. 2005; 11(20):7538–44.
8. Bonnett R. Photosensitizers of the porphyrin and phthalocyanine series for photodynamic therapy. Chem Soc Rev. 1995;24(1):19.
9. Frei KA, Bonel HM, Frick H, Walt H, Steiner RA. Photodynamic detection of diseased axillary sentinel lymph node after oral application of aminolevulinic acid in patients with breast cancer. Br J Cancer. 2004;90(4):805–9.
10. Kelty CJ, Brown NJ, Reed MWR, Ackroyd R. The use of 5-aminolaevulinic acid as a photosensitiser in photodynamic therapy and photodiagnosis. Photochem Photobiol Sci. 2002;1(3):158–68.
11. Knapp DW, Adams LG, Degrand AM, Niles JD, Ramos-Vara JA, Weil AB, et al. Sentinel lymph node mapping of invasive urinary bladder cancer in animal models using invisible light. Eur Urol. 2007;52(6):1700–8.
12. Ormond A, Freeman H. Dye sensitizers for photodynamic therapy. Materials (Basel). 2013; 6(3):817–40.
13. Dougherty TJ. An update on photodynamic therapy applications. J Clin Laser Med Surg. 2002;20(1):3–7.

14. Hillemanns P, Einstein MH, Iversen OE. Topical hexaminolevulinate photodynamic therapy for the treatment of persistent human papilloma virus infections and cervical intraepithelial neoplasia. Expert Opin Investig Drugs. 2015;24(2):273–81.
15. Helgesen ALO, Warloe T, Pripp AH, Kirschner R, Peng Q, Tanbo T, et al. Vulvovaginal photodynamic therapy vs. topical corticosteroids in genital erosive lichen planus: a randomized controlled trial. Br J Dermatol. 2015;173(5):1156–62.
16. Fingar VH, Kik PK, Haydon PS, Cerrito PB, Tseng M, Abang E, et al. Analysis of acute vascular damage after photodynamic therapy using benzoporphyrin derivative (BPD). Br J Cancer. 1999;79(11–12):1702–8.
17. Lui H, Hobbs L, Tope WD, Lee PK, Elmets C, Provost N, et al. Photodynamic therapy of multiple nonmelanoma skin cancers with verteporfin and red light-emitting diodes: two-year results evaluating tumor response and cosmetic outcomes. Arch Dermatol. 2004;140(1): 26–32.
18. Rousset N, Bourre L, Thibaud S. Sensitizers in photodynamic therapy. In: Thierry P, Royal Society of Chemistry, editors. Photodynamic therapy. Cambridge: Royal Society of Chemistry; 2003.
19. Triesscheijn M, Ruevekamp M, Aalders M, Baas P, Stewart FA. Outcome of mTHPC mediated photodynamic therapy is primarily determined by the vascular response. Photochem Photobiol. 2006;81(5):1161–7.
20. O'Connor AE, Gallagher WM, Byrne AT. Porphyrin and nonporphyrin photosensitizers in oncology: preclinical and clinical advances in photodynamic therapy. Photochem Photobiol. 2009;85(5):1053–74.
21. Zimcik P, Miletin M. Photodynamic therapy. In: Lang A, editor. Dyes and pigments: new research. New York: Nova; 2008.
22. Huang Z, Chen Q, Luck D, Beckers J, Wilson BC, Trncic N, et al. Studies of a vascular-acting photosensitizer, Pd-bacteriopheophorbide (Tookad), in normal canine prostate and spontaneous canine prostate cancer. Lasers Surg Med. 2005;36(5):390–7.
23. Huang Z. A review of progress in clinical photodynamic therapy. Technol Cancer Res Treat. 2005;4(3):283–93.
24. Schmidt MH, Bajic DM, Reichert KW, Martin TS, Meyer GA, Whelan HT. Light-emitting diodes as a light source for intraoperative photodynamic therapy. Neurosurgery. 1996; 38(3):552–6; discussion 556–7.
25. Chen J, Keltner L, Christophersen J, Zheng F, Krouse M, Singhal A, et al. New technology for deep light distribution in tissue for phototherapy. Cancer J. 2002;8(2):154–63.
26. Juzeniene A, Juzenas P, Ma L-W, Iani V, Moan J. Effectiveness of different light sources for 5-aminolevulinic acid photodynamic therapy. Lasers Med Sci. 2004;19(3):139–49.
27. Khan T, Unternährer M, Buchholz J, Kaser-Hotz B, Selm B, Rothmaier M, et al. Performance of a contact textile-based light diffuser for photodynamic therapy. Photodiagnosis Photodyn Ther. 2006;3(1):51–60.
28. Selm B, Rothmaier M, Camenzind M, Khan T, Walt H. Novel flexible light diffuser and irradiation properties for photodynamic therapy. J Biomed Opt. 2007;12(3):034024.
29. Buchholz J, Kaser-Hotz B. Kombination von photodynamischer Therapie und Hyperthermie mit wassergefiltertem Infrarotlicht zur Behandlung kutaner Plattenepithelkarzinome bei 15 Katzen: Eine Pilotstudie. Kleintierpraxis. 2010;50:248–54.
30. Korbelik M, Stott B, Sun J. Photodynamic therapy-generated vaccines: relevance of tumour cell death expression. Br J Cancer. 2007;97(10):1381–7.
31. Kelleher DK, Bastian J, Thews O, Vaupel P. Enhanced effects of aminolaevulinic acid-based photodynamic therapy through local hyperthermia in rat tumours. Br J Cancer. 2003;89(2): 405–11.
32. Onoyama M, Tsuka T, Imagawa T, Osaki T, Minami S, Azuma K, et al. Photodynamic hyperthermal chemotherapy with indocyanine green: a novel cancer therapy for 16 cases of malignant soft tissue sarcoma. J Vet Sci. 2014;15(1):117–23.
33. Gollnick SO, Vaughan L, Henderson BW. Generation of effective antitumor vaccines using photodynamic therapy. Cancer Res. 2002;62(6):1604–8.

34. Korbelik M, Sun J. Photodynamic therapy-generated vaccine for cancer therapy. Cancer Immunol Immunother. 2006;55(8):900–9.
35. Korbelik M, Cooper PD. Potentiation of photodynamic therapy of cancer by complement: the effect of gamma-inulin. Br J Cancer. 2007;96(1):67–72.
36. Dougherty TJ, Grindey GB, Fiel R, Weishaupt KR, Boyle DG. Photoradiation therapy. II. Cure of animal tumors with hematoporphyrin and light. J Natl Cancer Inst. 1975;55(1):115–21.
37. Lucroy MD. Photodynamic therapy for companion animals with cancer. Vet Clin North Am Small Anim Pract. 2002;32(3):693–702, viii.
38. Buchholz J, Walt H. Veterinary photodynamic therapy: a review. Photodiagnosis Photodyn Ther. 2013;10(4):342–7.
39. Magne ML, Rodriguez CO, Autry SA, Edwards BF, Theon AP, Madewell BR. Photodynamic therapy of facial squamous cell carcinoma in cats using a new photosensitizer. Lasers Surg Med. 1997;20(2):202–9.
40. McCaw DL, Pope ER, Payne JT, West MK, Tompson RV, Tate D, et al. Treatment of canine oral squamous cell carcinomas with photodynamic therapy. Br J Cancer. 2000;82(7):1297–9.
41. McCaw DL, Payne JT, Pope ER, West MK, Tompson RV, Tate D. Treatment of canine hemangiopericytomas with photodynamic therapy. Lasers Surg Med. 2001;29(1):23–6.
42. Hahn K, Panjehpour M, Legendre AM. Photodynamic therapy response in cats with cutaneous squamous cell carcinoma as a function of fluence. Vet Dermatol. 1998;9:3–7.
43. Buchholz J, Walt H, Fidel J, Al E. Photodynamische Therapie beim felinen Plattenepithelkarzinom sowie bei versch. kaninen Tumoren. Kleintierpraxis. 2003;48: 405–18.
44. Buchholz J, Wergin M, Walt H, Gräfe S, Rohrer Bley C, Kaser-Hotz B. Photodynamic therapy of feline cutaneous squamous cell carcinoma using a newly developed liposomal photosensitizer: preliminary results concerning drug safety and efficacy. J Vet Intern Med. 2007;21(4):770.
45. Borgatti-Jeffreys A, Hooser SB, Miller MA, Lucroy MD. Phase I clinical trial of the use of zinc phthalocyanine tetrasulfonate as a photosensitizer for photodynamic therapy in dogs. Am J Vet Res. 2007;68(4):399–404.
46. Chang SC, Buonaccorsi G, MacRobert A, Bown SG. Interstitial and transurethral photodynamic therapy of the canine prostate using meso-tetra-(m-hydroxyphenyl) chlorin. Int J Cancer. 1996;67(4):555–62.
47. Chang SC, Chern IF, Hsu YH. Biological responses of dog prostate and adjacent structures after meso-tetra-(m-hydroxyphenyl) chlorin and aluminum disulfonated phthalocyanine based photodynamic therapy. Proc Natl Sci Counc Repub China B. 1999;23(4):158–66.
48. Nseyo UO, DeHaven J, Dougherty TJ, Potter WR, Merrill DL, Lundahl SL, et al. Photodynamic therapy (PDT) in the treatment of patients with resistant superficial bladder cancer: a long-term experience. J Clin Laser Med Surg. 1998;16(1):61–8.
49. Anderson TM, Dougherty TJ, Tan D, Sumlin A, Schlossin JM, Kanter PM. Photodynamic therapy for sarcoma pulmonary metastases: a preclinical toxicity study. Anticancer Res. 2003;23(5A):3713–8.
50. Gloi AM, Beck E. Threshold dose of three photosensitizers in dogs with spontaneous tumors. Vet Ther. 2003;4(3):269–78.
51. Gloi AM, Beck E. Evaluation of porfimer sodium in dogs and cats with spontaneous tumors. Vet Ther. 2004;5(1):26–33.
52. Du KL, Mick R, Busch TM, Zhu TC, Finlay JC, Yu G, et al. Preliminary results of interstitial motexafin lutetium-mediated PDT for prostate cancer. Lasers Surg Med. 2006;38(5):427–34.
53. Ross HM, Smelstoys JA, Davis GJ, Kapatkin AS, Del Piero F, Reineke E, et al. Photodynamic therapy with motexafin lutetium for rectal cancer: a preclinical model in the dog. J Surg Res. 2006;135(2):323–30.
54. Osaki T, Hoshino S, Hoshino Y, Takagi S, Okumura M, Kadosawa T, et al. Clinical pharmacokinetics of anti-angiogenic photodynamic therapy with benzoporphyrin derivative monoacid ring-a in dogs having naturally occurring neoplasms. J Vet Med A Physiol Pathol Clin Med. 2006;53(2):108–12.

55. Xiao Z, Dickey D, Owen RJ, Tulip J, Moore R. Interstitial photodynamic therapy of the canine prostate using intra-arterial administration of photosensitizer and computerized pulsed light delivery. J Urol. 2007;178(1):308–13.
56. Liu Y, Ma XQ, Jin P, Li HT, Zhang RR, Ren XL, et al. Apoptosis induced by hematoporphyrin monomethyl ether combined with He-Ne laser irradiation in vitro on canine breast cancer cells. Vet J. 2011;188(3):325–30.
57. Li HT, Song XY, Yang C, Li Q, Tang D, Tian WR, et al. Effect of hematoporphyrin monomethyl ether-mediated PDT on the mitochondria of canine breast cancer cells. Photodiagnosis Photodyn Ther. 2013;10(4):414–21.
58. Rocha MST, Lucci CM, Longo JPF, Galera PD, Simioni AR, Lacava ZGM, et al. Aluminum-chloride-phthalocyanine encapsulated in liposomes: activity against naturally occurring dog breast cancer cells. J Biomed Nanotechnol. 2012;8(2):251–7.
59. Lin N, Li C, Wang Z, Zhang J, Ye X, Gao W, et al. A safety study of a novel photosensitizer, sinoporphyrin sodium, for photodynamic therapy in Beagle dogs. Photochem Photobiol Sci. 2015;14(4):815–32.
60. Lucroy MD, Edwards BF, Peavy GM, Krasieva TB, Griffey SM, Stiles JB, et al. Preclinical study in cats of the pro-photosensitizer 5-aminolevulinic acid. Am J Vet Res. 1999;60(11):1364–70.
61. Ridgway TD, Lucroy MD. Phototoxic effects of 635-nm light on canine transitional cell carcinoma cells incubated with 5-aminolevulinic acid. Am J Vet Res. 2003;64(2):131–6.
62. Lucroy MD, Ridgway TD, Peavy GM, Krasieva TB, Higbee RG, Campbell GA, et al. Preclinical evaluation of 5-aminolevulinic acid-based photodynamic therapy for canine transitional cell carcinoma. Vet Comp Oncol. 2003;1(2):76–85.
63. Huang Z, Chen Q, Trncic N, LaRue SM, Brun P-H, Wilson BC, et al. Effects of Pd-bacteriopheophorbide (TOOKAD)-mediated photodynamic therapy on canine prostate pretreated with ionizing radiation. Radiat Res. 2004;161(6):723–31.
64. L'eplattenier HF, Klem B, Teske E, van Sluijs FJ, van Nimwegen SA, Kirpensteijn J. Preliminary results of intraoperative photodynamic therapy with 5-aminolevulinic acid in dogs with prostate carcinoma. Vet J. 2008;178(2):202–7.

Chapter 10
Clinical Applications of Cancer PDT

Julia Buchholz

Abstract Still today photodynamic therapy (PDT) represents a little known treatment modality in veterinary oncology even though there are some evident advantages especially for the treatment of animals: PDT can usually be done as an outpatient or day-case setting, it is convenient for the patient since usually one treatment is sufficient, and it has few to no side effects. PDT has several potential advantages over other local treatment modalities such as surgery and radiotherapy (RT): it is comparatively noninvasive, repeated doses can be given without the total-dose limitations associated with RT, and the healing process results in little or no scarring. In human oncology, PDT has been introduced more than 35 years ago, and since about 10 years, it is used more widely in the clinic. The treatment itself is quite easily feasible even though the PDT biology behind is rather complex. Clinical studies conducted in veterinary medicine describe the use of different photosensitizers applied systemically or topically as well as different systems of light delivery. These studies are presented in the current chapter. In veterinary oncology, the main indication is represented by the most common skin neoplasia in cats: cutaneous squamous cell carcinoma. Results are very promising. In the future, other indications might become treatable with PDT, especially considering the evolution of more selective photosensitizing drugs and newer, less expensive light systems. Possible indications include, but are not limited to, superficial canine transitional carcinoma, equine sarcoids, and equine squamous cell carcinoma as well as neoplasia in exotic animals.

10.1 Introduction

There are several advantages using PDT, especially in veterinary medicine, such as: not being invasive, single treatment on outpatient basis, repeatable if necessary, uncomplicated healing with very little scar formation and possible combination with other treatment modalities. PDT might be combined with surgery, radiation

J. Buchholz
Animal Oncology and Imaging Center, Huenenberg, Switzerland
e-mail: buchholz@aoicenter.ch

F.P. Sellera et al. (eds.), *Photodynamic Therapy in Veterinary Medicine: From Basics to Clinical Practice*, DOI 10.1007/978-3-319-45007-0_10

therapy, hyperthermia, immunotherapy (as discussed in chapter 9) and chemotherapy. Problems known from other therapies commonly used in oncology, such as resistant tumor cells (especially using chemotherapy but also known for radiotherapy) have been described against photosensitizers as well [1]. This problem represents an important rationale for combining different treatment modalities, especially for therapy-resistant cases.

10.2 Companion Animals

10.2.1 Squamous cell carcinoma

Squamous cell carcinomas (SCC) are frequently occurring neoplasms in veterinary patients, mainly in cats and less often in dogs. This tumor accounts for approximately 15 % of all cutaneous tumors in the cat. There is a predisposition for nonpigmented animals or nonpigmented area, and an etiologic correlation between development of the neoplasm and exposure to ultraviolet light does exist. The most common cutaneous locations for cats are the sparsely haired area of the nasal planum, eyelids, and pinnae. Generally, squamous cell carcinoma is locally invasive but late to metastasize [2]. The established standard therapies are surgery and radiation therapy. Several studies have shown the efficacy of PDT to treat feline squamous cell carcinoma [3–9].

Feline superficial SCC/carcinoma in situ still represents the main indication in veterinary medicine.

Bowen's disease represents an own entity defined as SCC in situ (not penetrating the basal membrane), but although being superficial, they are often more difficult to control with PDT and they can cause multiple lesions.

In a study by Magne et al., 51 cats with cutaneous SCC received the photosensitizer pyropheophorbid-alpha-hexyl-ether (HPPH-23) intravenously. Forty-nine percent of the cats showed a complete tumor remission, 12 % showed partial tumor remission, and 39 % did not respond to PDT. There was a significant correlation between complete remission and length of local tumor control with tumor stage: the smaller and less invasive the tumors were, the better they responded to PDT [8].

In another study including only six cats treated with EtNBS as photosensitizer, two partial and four complete remissions could be observed [10].

Stell et al. achieved a complete remission rate of 85 % using topical ALA-PDT for superficial feline SCC. As light source, they used a light-emitting diode (LED) instead of a laser. After a median time of 21 weeks, they had a recurrence rate of 64 % though [6]. In a second study using the same protocol, a recurrence could be seen in 51 % of the cats having a complete response after PDT with a median time to recurrence of 157 days [3].

In another study also using LEDs as light source and the hematoporphyrin derivative Photogem® as photosensitizer for small noninvasive tumors, satisfactory results could be obtained [11].

Reeds et al. used the photosensitizer HPPH with an LED to treat noninvasive carcinoma in dogs and cats. Eight out of the nine treated tumors showed a complete remission, and >50 % of the patients did not show tumor recurrence for the follow-up time of 68 weeks [12].

Due to the generally unsatisfactory results while treating more invasive tumor stages, a study was conducted to find out if increasing the dose would ameliorate the results treating invasive stages. Cats with invasive SCC, treated with PDT using aluminum phthalocyanine tetrasulfonate ($AlPcS_4$) as photosensitizer, showed a significantly shorter time of median remission using 100 J/cm^2 ($n=8$; 69 days; range 0–619 days) than using 200 J/cm^2 ($n=6$; 522 days; range 151–1,057 days) [13].

The author of this article is most experienced using an intravenously administered photosensitizer (*m*THPC). *m*THPC is a very potent photosensitizer that is approved for the treatment of head and neck cancer in humans. It requires very low drug and light doses to be effective. The substance is a single component of 98 % purity.

In one study, we used a conventional lipophilic formulation of *m*THPC (Foscan) and a new liposomal formulation (Fospeg®) to determine their pharmacokinetic behavior. In ten cats, in vivo fluorescence intensity measurements of tumor and skin were performed with a fiber spectrophotometer after intravenous injection of *m*THPC (either Foscan® or Fospeg). Blood samples, drawn at several time points after photosensitizer administration, were analyzed by HPLC to investigate the differences within the plasma compartment. Organ parameters were analyzed serially in all cats and VEGF levels were determined prior to PDT.

Fluorescence intensities, fluorescence ratios (tumor fluorescence divided by skin fluorescence), and bioavailability in the tumor were two to four times higher with Fospeg® compared to Foscan®. Fospeg concentration in the tumor increased constantly to reach a maximum at 4 h after injection. Plasma concentration and plasma bioavailability were about three times higher with Fospeg compared to Foscan® measured at the time points of highest plasma concentration. We also performed power Doppler ultrasound (PDUS) before and at several time points after treatment to investigate the vascular effects of PDT on tumors. Images were digitized for assessment of vascularity and blood volume. On PDUS, a significant decrease in vascularity and blood volume was noted after PDT. Lowest values were found 24 h after PDT. The distinct effect on tumor vessels shown by PDUS indicates the importance of the vascular component in tumor destruction with *m*THPC-PDT. Stage, tumor location, pre-PDT VEGF levels, and presence of side effects could not be established as prognostic indicators.

All cats responded to therapy, with a complete response rate of 93 %. The tumor recurrence rate was 39 % with a median time to recurrence of about 18 months.

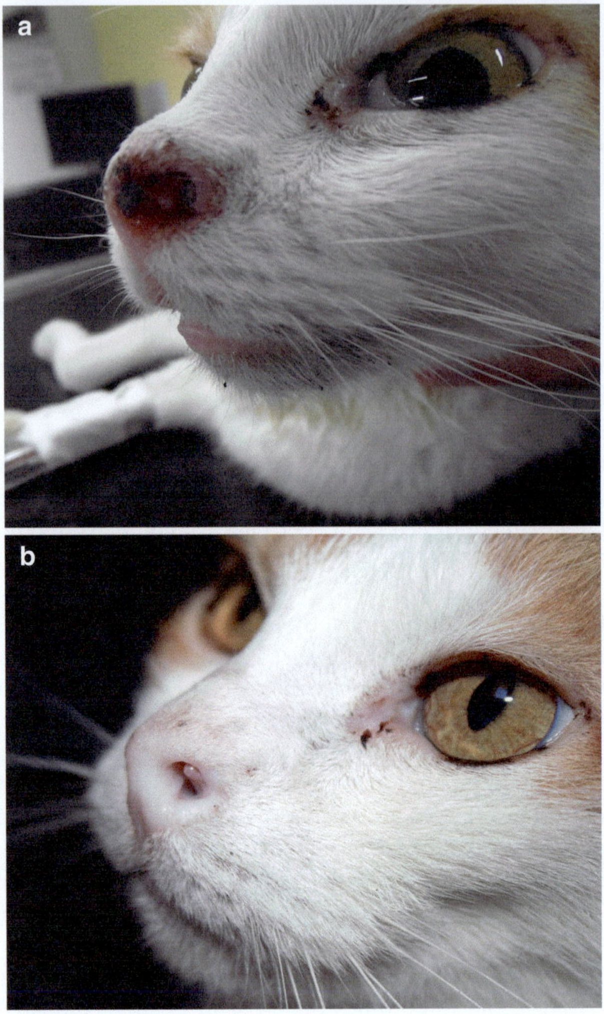

Fig. 10.1 Cat with nasal squamous cell carcinoma. (**a**) Before PDT; (**b**) After PDT

Local control could be achieved in 62 % of patients at 1 year post therapy. Eighty-three percent of the tumors treated with the lipophilic formulation and 27 % of the tumors treated with the liposomal formulation showed tumor recurrence. We concluded that the favorable pharmacokinetics of the liposomal drug seem to translate into superior tumor control [14] (Fig. 10.1).

In a subsequent study, we treated 20 cats with the liposomal formulation only. Following the results of the pharmacokinetic study, cats were treated with laser light (wavelength of 652 nm) 4–6 h after injection (Fig. 10.2). The tissues surrounding the area to be treated should be covered (thick black tape is sufficient (Fig. 10.3a) and the eye has to be protected (with wax or wax-covered gold shields) if it is in/

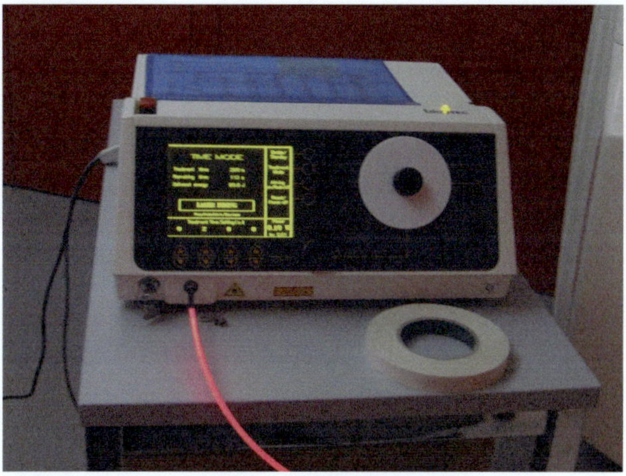

Fig. 10.2 Laser device used in PDT procedures. Note the optical fiber coupled to the laser output

near the treatment field (Fig. 10.3b). All patients have been treated with a front lens device (Fig. 10.4a) even though we have some preliminary results (not included within this study) using interstitial illumination (Fig. 10.4b).

The animals had to be anesthetized to keep still for the entire laser light application (about 3 min). The patients showed an initial complete response rate of 100 % and tumor control in 75 % after 1 year. The recurrence rate was 20 % with a median time to recurrence of 172 days. The intravenous administration was well tolerated by all cats (Fig. 10.5a). Due to the high selectivity of the liposomal formulation of the photosensitizer, the generalized light sensitivity of the patient was only about 10–14 days [5]. Local application of the photosensitizer (Fig. 10.5b) does not cause generalized photosensitivity, and therefore, patients can go outside immediately after treatment. Animal patients might try to lick/scratch the locally administered substances off.

10.2.2 Carcinoma of the Lower Urinary Tract

In human medicine, PDT has been successfully used for bladder carcinoma [16, 17]. Also in veterinary medicine, especially transitional cell carcinoma of the urinary bladder may represent a promising indication, at least for initial stages.

In the basic studies section, some of the published work in veterinary medicine has already been presented. Five dogs with transitional cell carcinoma of the urinary bladder [18] and one dog with a prostatic carcinoma [18] showed transient improvement of clinical symptoms. In the first study, dogs with histological or cytological confirmation of TCC of the lower urinary tract were given 60 mg/kg of ALA orally, and 4 h later they were treated with 635 nm laser light using a laser fiber delivery system (Pioneer Optics, Windsor Lock, CT, USA). This system was passed retrograde through

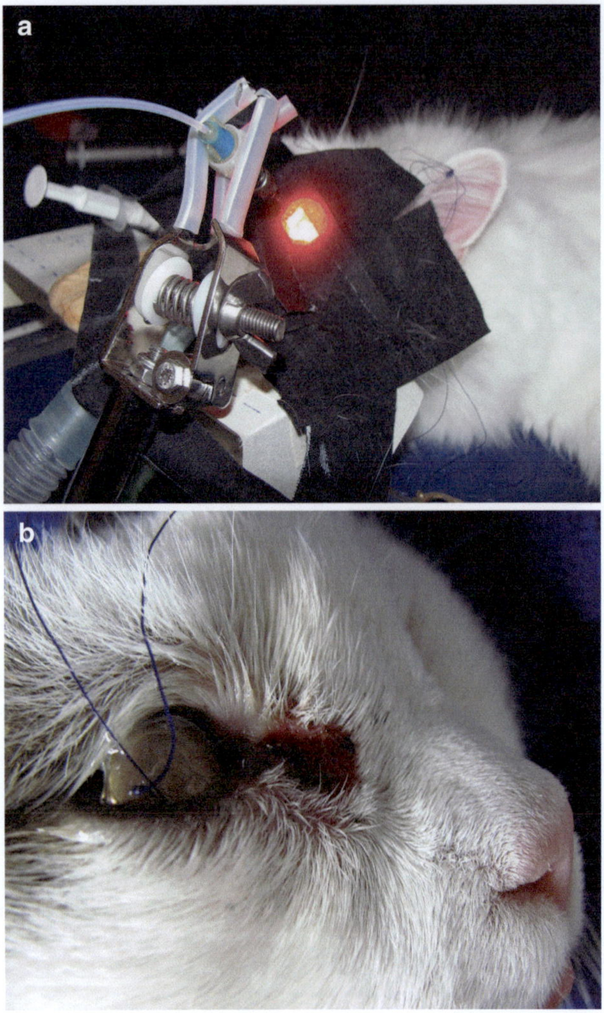

Fig. 10.3 (**a**) Cat with squamous cell carcinoma on medial canthus of the right eyelid. (**a**) In general anesthesia during PDT; (**b**) Gold shield in place to protect the eye during PDT

the urethra into the urinary bladder. The fiber delivery system consisted of a 400-mm diameter optical fiber that terminated in a 1-cm cylindrical diffuser. The fiber tip was visualized using ultrasonography. The energy density was 100 J/cm^2 and the power density was 75 mW/cm^2. The treatment time was 22 min and 13 s. The first dog treated had preexisting hydronephrosis and hydroureter and became anuric within 12 h of PDT. Due to deterioration, the dog was euthanized 72 h after PDT. For the remaining five dogs, the median progression-free interval (PFI) was 6 weeks (range 4–34 weeks). Possible side effects after PDT were stranguria and hematuria. So far there is no well-established protocol available leading to longer-term control, and to the authors' knowledge, PDT is not yet routinely used clinically for this indication.

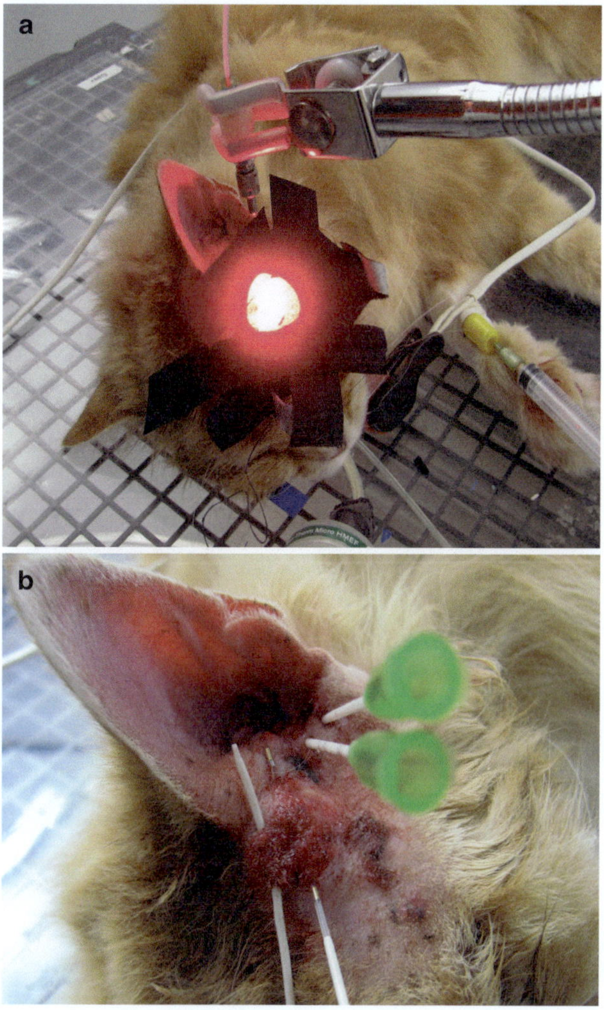

Fig. 10.4 (**a**) Cat in general anesthesia positioned during PDT for a squamous cell carcinoma on the left temple (superficial lesion); (**b**) The same cat in preparation for interstitial PDT for a bulky lesion at the left ear base – the light fiber is inserted into the second catheter from the left

10.2.3 Miscellaneous Tumors

In 1991, PDT was successfully used for various neoplasms in dogs, cats, and snakes. The second-generation photosensitizer chloroaluminum sulfonated phthalocyanine (CASPc) was given IV 48 h prior to irradiation with 674 nm light to ten cats, two dogs, and three snakes with various tumor types. Some larger tumors were debulked surgically prior to PDT. A 67 % complete response rate and 22 %

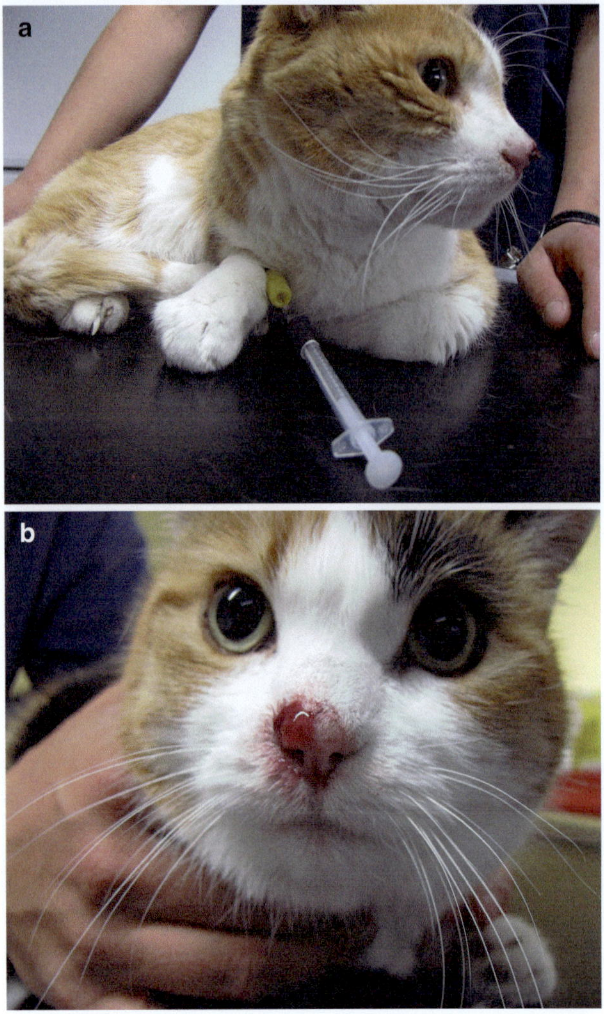

Fig. 10.5 (**a**) Cat during intravenous injection of the photosensitizer – the photosensitizer is protected from light by covering the drug containing part of the syringe with a black tape; (**b**) Cat with the topically applied photosensitizer on the nasal planum

partial response rate were observed, and 11 % of tumors did not respond to the treatment (see Sect. 8.3.2?).

Payne et al. had promising results while treating canine hemangiopericytoma and canine oral SCC with a combination of PDT and surgery [19]. In a subsequent work, the same authors confirmed these results for oral SCC [20], whereas the results for hemangiopericytoma could not be reproduced [21]. Nine of 11 dogs with oral SCC treated with HPPH-PDT, partly in combination with surgery, showed a complete tumor remission for at least 17 months. Eight dogs were considered as healed; the median survival time was 29 months. The cosmetic outcome was described as very good [20].

In a study, 14 dogs with oral and nasal tumors were treated with antivascular PDT using benzoporphyrin derivate monoacid ring A (BPD-MA). Contrast-enhanced tumors were observed before antivascular PDT, whereas they no longer enhanced after PDT. The authors conclude that PDT might be an option for patients that cannot be treated with current antitumor therapies (radiotherapy or surgery). The 1-year survival rate in dogs with oral tumors and nasal tumors was 71 % and 57 %, respectively [22].

Lucroy et al. could achieve a temporary improvement of clinical symptoms in three dogs and one cat with tumors of the nasal cavity [15] – radiation therapy, if available, is usually considered treatment of choice for this tumor, being able to control the tumor for a longer period of time. A case report describes PDT for a SCC of the esophagus in a dog with the result of partial remission [23]. Treatment approaches for feline oral SCC, scleral melanoma, and canine mast cell tumors are described where EtNBS was used as a photosensitizer and generally 400 J of 652 nm light was given at a short drug-light interval of only 3 h. Pharmacokinetics have not been performed and also timing of optimized photoirradiation or optimal fluence rate was not known. All animals were exposed to normal daylight after less than 5 days without residual photosensitization. They conclude that EtNBS-PDT is safe for dogs and cats and that it has activity against selected naturally occurring tumors with an overall objective response rate of 61.5 % [10]. A study from 2004 resulted in complete remission in two mast cell tumors in dogs and of one basal cell carcinoma in a cat using PAD-S31 as photosensitizer. The photosensitizer was administered IV and the animals showed no cutaneous photosensitivity under room light illumination. Further investigation is needed according to the authors [24].

PDT as an adjunct to treat osseous tumors was considered in a study of seven dogs with spontaneous osteosarcoma of the distal radius, and it could be shown that PDT is able to penetrate relatively large osseous tumors [25].

Obviously there are many reports about PDT in laboratory mice and rats having induced or transplanted tumors. Even though these studies are not primarily looking at long-term tumor control or drug safety, some of these data could be helpful to establish protocols treating small mammals. Lucroy described the treatment of a SCC in a ferret with PDT showing a short-term partial response [26].

10.2.4 Potential Side Effects

After intravenous application of ALA in cats, hematological toxicity and hepatotoxicity were observed [7] – this could not be observed with topical administration. In dogs, oral administration of ALA caused vomiting in 70 % of patients [15]. After intravenous administration of the lipophilic formulation of *m*THPC (Foscan®), systemic side effects such as tachypnea and tachycardia during injection of the drug could be observed [14]. During injection of EtNBS, systemic reactions such as nausea and vomiting could be observed [10].

With most of the new photosensitizers, systemic side effects are rarely seen. During laser light administration, pain sensation is described in humans and increased heart rates can be observed (probably caused by pain sensation) in animals. Local side effects such as hyperemia, edema, cyanosis, and pruritus in the treated area have been observed [5, 6, 8] from hours to weeks after PDT (Fig. 10.6).

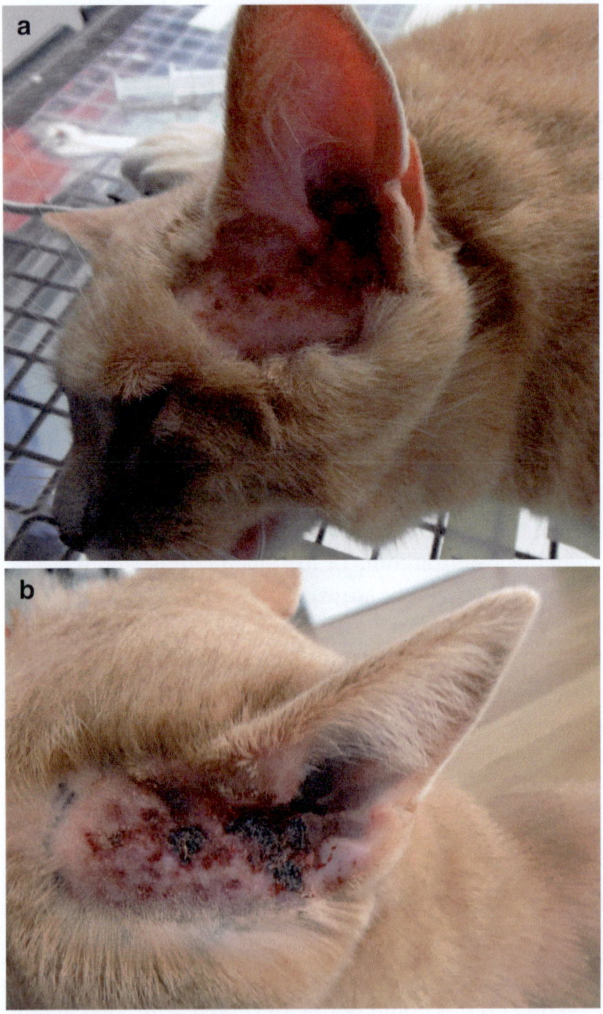

Fig. 10.6 Cat with squamous cell carcinoma on the left temple area. (**a**) Before PDT; (**b**) Local edema formation in the cat 4 days after PDT

Depending on the size of the lesion and degree of associated inflammation, patients usually will be treated with systemic anti-inflammatory agents and antibiotics for 2 weeks after PDT. In most cases, a crust (Fig. 10.7a) will develop within the treated region, which will remain for several weeks (depending on the depth of the lesion) and then will shed off. The skin and mucous membrane will re-epithelialize afterward and the hair will grow back (Fig. 10.7b). The paravenous injection of *m*THPC can cause local irritation and increased photosensitivity at that site (Fig. 10.8a). Also, the skin usually remains thinner for a few months before getting back to

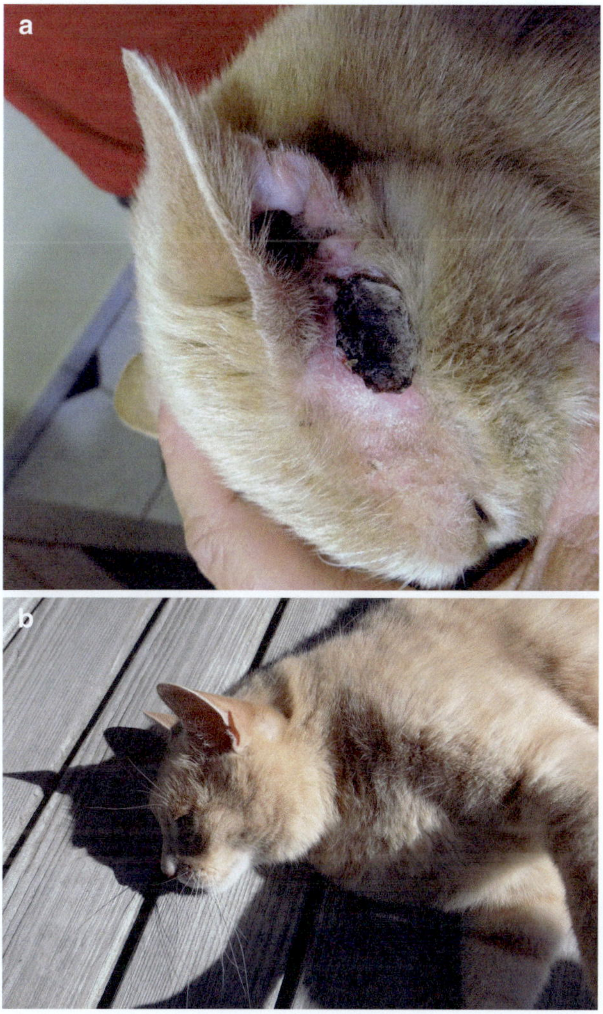

Fig. 10.7 Cat with squamous cell carcinoma on the left temple area. (**a**) Crust still remains in place about 6 weeks after PDT; (**b**) Normal healing and hair growth with tumor in complete remission 6 months post PDT

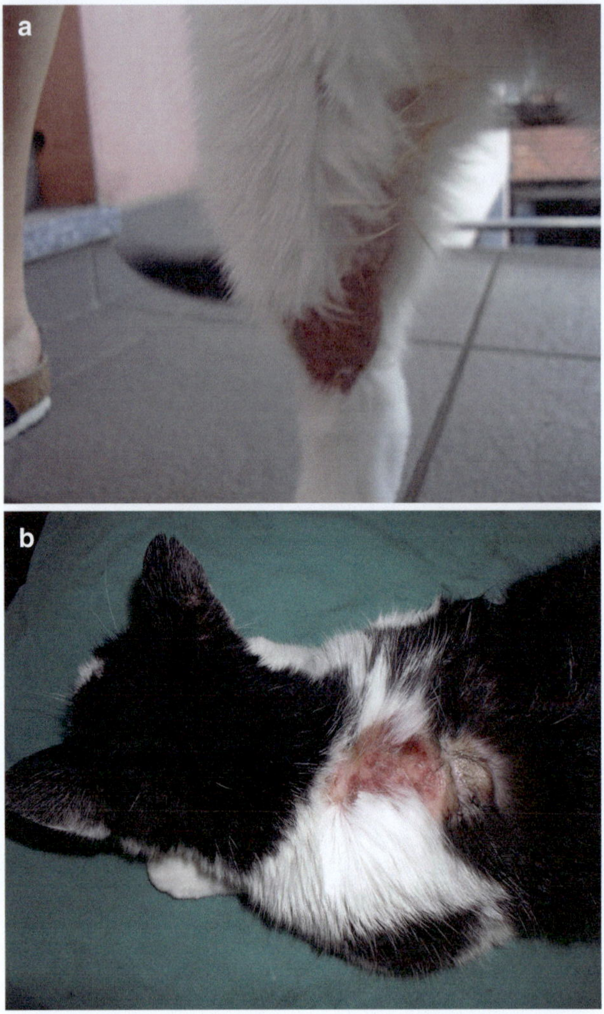

Fig. 10.8 (**a**) Cat with severe reaction at the site of intravenous photosensitizer injection – partly paravenous injection might have occurred. (**b**) The same cat presenting alopecia and moist dermatitis between the shoulder blades that went outside during the day few days after systemic PDT – the reactions are more severe in the white-haired area

normal. If the animals go outside during daylight in the phase of generalized photosensitivity, they might present with sunburn-like changes on the skin and the overlying hair will fall off (Fig. 10.8b) as well as swelling of the head, eyelids, etc.

10.3 Farm Animals

In large animals, cancer is far less common compared to small animals, but there are few reports using PDT for large animal neoplasia.

10.3.1 Equines

10.3.1.1 Periocular Squamous Cell Carcinoma (PSCC)

Giuliano et al. are working on the inclusion of PDT for the treatment of equine periocular squamous cell carcinoma (PSCC). In 2008 they published a pilot study using HPPH locally after surgical resection, and they concluded that PDT could be considered a safe and effective novel treatment for PSCC in horses but that further research is needed [27]. Prior to a second clinical study, the authors looked into the cellular localization of verteporfin as a function of time after local injection in an in vivo model of squamous cell carcinoma [28]. In their second clinical study, they used verteporfin as a photosensitizer and compared horses treated by surgery and cryotherapy (group 1; $n=14$) with horses treated through surgery and PDT (group 2; $n=10$). Signalment, tumor laterality, and size were not significantly different between groups. Time to recurrence was significantly different between groups ($p=0.0006$). In group 1, 11/14 horses had tumor regrowth with a median time to recurrence of 10 months. In group 2, no horse demonstrated tumor recurrence after one treatment with excision and PDT (minimum follow-up of 25 months). The authors therefore concluded that the likelihood of tumor recurrence following surgery in equine PSCC was significantly reduced with local verteporfin PDT compared with cryotherapy [29].

10.3.1.2 Sarcoids

Equine sarcoids are commonly occurring tumors in horses. These tumors do not metastasize, but are very difficult to control locally, therefore presenting a challenge for therapy. In addition, these tumors are often localized in delicate areas where extensive surgery is not possible (e.g., periocular, nostril, inguinal area). Common therapy modalities include surgery, cryosurgery, laser surgery, immunotherapy, chemotherapy (cisplatin), and radiation therapy (radioactive iridium); however, none of them show satisfying success. Preclinical studies have shown the efficacy of PDT to treat equine sarcoids. In vitro cytotoxicity and in vivo antitumoral action could be shown using hypericin. Four intratumoral injections were given to the in vivo subject, a donkey with three equine sarcoids, with daily illumination for 25 days. At the end of therapy, an 81 % tumor reduction could be observed and 2 months later a 90 % reduction was observed [30]. A clinical study, where a photosensitizer was applied, topically showed efficacy, too (Gustafson 2001, data not published). A case report showed good results using intralesional *m*THPC +/– surgery [31]. For very large sarcoids, debulking surgery prior to PDT might be indicated. Also, equine skin itself is relatively thick and might be pigmented making it much harder for the light to penetrate into depth compared to feline or canine skin. Equine sarcoids are considered immunogenic tumors, and therefore, if PDT would be able to have an influence on the patients' immune system, this might improve response rates. The introduction of PDT and vaccines to further stimulate an immunogenic response might become important in the future especially for this tumor type. Intralesional

PDT would be preferable to avoid generalized photosensitivity and to avoid the necessity of large amounts of photosensitizers due the body weight of horses. For liposomal *m*THPC, it could be shown in a mouse model that the fluorescence maximum was reached at 24 h post-intratumoral injection and this time point correlated with the best therapeutic effect [32].

10.3.2 Cattle

There is one case report describing the treatment of a cow with an ocular squamous cell carcinoma involving the third eyelid and conjunctiva with intratumoral 5-ALA-PDT. After LED irradiation with a wavelength of 600–700 nm, the tumor size decreased to 50 % after 15 days, and there was a complete remission after 3 months. The patient was tumor-free for 12 months post treatment [33].

10.4 Exotic, Zoo, and Wildlife Animals

10.4.1 Birds

A case report describes PDT of a SCC in the casque of a great hornbill (*Buceros bicornis*). HPPH was used IV and after 24 h the tumor was illuminated with a 665 nm diode laser. After treatment, the bird was kept indoors for 3 weeks to prevent damage due to photosensitivity. They could observe some tumor necrosis, but after few weeks, the tumor continued to grow and a second PDT was performed after tumor debulking. During anesthesia, the bird went into respiratory distress but then recovered. The tumor could not be controlled though, probably due to tumor size and bone invasion [34].

The same group described PDT of a patagial SCC in an African rose-ringed parakeet. Unfortunately also in this bird, PDT could not be shown to be successful. The authors conclude that the PDT protocols in birds might have to be designed differently to those used in other animals [34].

Another report describes porfimer sodium-based PDT of a SCC in a cockatiel. The tumor had a partial response after each of several PDT sessions but recurred quickly after treatment. Edema of the hooks was described several weeks after PDT, most likely due to cutaneous photosensitivity [35].

Further study with regard to the utility of PDT in avian medicine is needed, but since cutaneous cancer seems to be most common in birds, there is potential to further try to improve PDT for birds.

10.4.2 Reptiles

Three snakes have been treated with chloroaluminum sulfonated phthalocyanine-based PDT. The photosensitizer was given into the palatine vein of the caudal venous sinus 48 h before light exposure. An oral SCC in a pet *Boa constrictor* had a complete remission followed by tumor recurrence 28 weeks after PDT. A second PDT resulted in a long-term complete remission. A zoo-kept Burmese python with a mixed tumor of the palatine gland had a partial remission with tumor recurrence 6 weeks after PDT. A second PDT caused tumor necrosis but soon after the snake died because of bacterial pneumonia. A zoo-kept European viper with an adenocarcinoma of the cloaca died 4 days after PDT. A possible reason could be tumor extension into the oviduct [36]. There is another retrospective study of a boa constrictor showing a short-term response to PDT [37].

References

1. Casas A, Perotti C, Di Venosa G, Batlle A. Mechanisms of resistance to photodynamic Therapy: an update. In: Resistance to photodynamic therapy in cancer. 2015; p. 29–63.
2. Hauck M. Tumors of the skin and subcutaneous tissues. In: Withrow & MacEwen's small animal clinical oncology. 5th ed. Missouri: Elsevier; 2013. p. 310–2.
3. Bexfield NH, Stella J, Gear RN, Dobson JM. Photodynamic therapy of superficial nasal planum squamous cell carcinomas in cats: 55 cases. J Vet Intern Med. 2008;22(6):1385–9.
4. Buchholz J, Walt H, Fidel J, Al E. Photodynamische therapie beim felinen plattenepithelkarzinom sowie bei versch. kaninen tumoren. Kleintierpraxis. 2003;48:405–18.
5. Buchholz J, Wergin M, Walt H, Gräfe S, Rohrer Bley C, Kaser-Hotz B. Photodynamic therapy of feline cutaneous squamous cell carcinoma using a newly developed liposomal photosensitizer: preliminary results concerning drug safety and efficacy. J Vet Intern Med. 2007;21(4):770.
6. Stell AJ, Dobson JM, Langmack K. Photodynamic therapy of feline superficial squamous cell carcinoma using topical 5-aminolaevulinic acid. J Small Anim Pract. 2001;42(4):164–9.
7. Lucroy MD, Edwards BF, Peavy GM, Krasieva TB, Griffey SM, Stiles JB, et al. Preclinical study in cats of the pro-photosensitizer 5-aminolevulinic acid. Am J Vet Res. 1999;60(11):1364–70.
8. Magne ML, Rodriguez CO, Autry SA, Edwards BF, Theon AP, Madewell BR. Photodynamic therapy of facial squamous cell carcinoma in cats using a new photosensitizer. Lasers Surg Med. 1997;20(2):202–9.
9. Peaston AE, Leach MW, Higgins RJ. Photodynamic therapy for nasal and aural squamous cell carcinoma in cats. J Am Vet Med Assoc. 1993;202(8):1261–5.
10. Frimberger AE, Moore AS, Cincotta L, Cotter SM, Foley JW. Photodynamic therapy of naturally occurring tumors in animals using a novel benzophenothiazine photosensitizer. Clin Cancer Res. 1998;4(9):2207–18.
11. Ferreira I, Rahal SC, Rocha NS, Gouveia AH, Corrêa TP, Carvalho YK, et al. Hematoporphyrin-based photodynamic therapy for cutaneous squamous cell carcinoma in cats. Vet Dermatol. 2009;20(3):174–8.

12. Reeds KB, Ridgway TD, Higbee RG, Lucroy MD. Non-coherent light for photodynamic therapy of superficial tumours in animals. Vet Comp Oncol. 2004;2(3):157–63.
13. Hahn KA, Panjehpour M, Legendre AM. Photodynamic therapy response in cats with cutaneous squamous cell carcinoma as a function of fluence. Vet Dermatol. 1998;9:3–7.
14. Buchholz J, Kaser-Hotz B, Khan T, Rohrer Bley C, Melzer K, Schwendener RA, et al. Optimizing photodynamic therapy: in vivo pharmacokinetics of liposomal meta-(tetrahydroxyphenyl) chlorin in feline squamous cell carcinoma. Clin Cancer Res. 2005;11(20):7538–44.
15. Lucroy MD, Ridgway TD, Peavy GM, Krasieva TB, Higbee RG, Campbell GA, et al. Preclinical evaluation of 5-aminolevulinic acid-based photodynamic therapy for canine transitional cell carcinoma. Vet Comp Oncol. 2003;1(2):76–85.
16. Nseyo UO, Merrill DC, Lundahl SL. Green light photodynamic therapy in the human bladder. Clin Laser Mon. 1993;11(5):247–50.
17. Nseyo UO, DeHaven J, Dougherty TJ, Potter WR, Merrill DL, Lundahl SL, et al. Photodynamic therapy (PDT) in the treatment of patients with resistant superficial bladder cancer: a long-term experience. J Clin Laser Med Surg. 1998;16(1):61–8.
18. Ridgway TD, Lucroy MD. Phototoxic effects of 635-nm light on canine transitional cell carcinoma cells incubated with 5-aminolevulinic acid. Am J Vet Res. 2003;64(2):131–6.
19. Payne JT, McCaw DL, Rogers KJ, Tompson RV. Treatment of naturally occurring hemangiopericytoma and oral squamous cell carcinoma in dogs using surgery and photodynamic therapy with HPPH as a photosensitizer. RR, Watson GM, Steiner RW, Johnson DE, Shapshay SM, Pankratov MM, et al. editors. SPIE; 1995;2395:622–6.
20. McCaw DL, Pope ER, Payne JT, West MK, Tompson RV, Tate D, et al. Treatment of canine oral squamous cell carcinomas with photodynamic therapy. Br J Cancer. 2000;82(7):1297–9.
21. McCaw DL, Payne JT, Pope ER, West MK, Tompson RV, Tate D. Treatment of canine hemangiopericytomas with photodynamic therapy. Lasers Surg Med. 2001;29(1):23–6.
22. Osaki T, Takagi S, Hoshino Y, Okumura M, Kadosawa T, Fujinaga T. Efficacy of antivascular photodynamic therapy using benzoporphyrin derivative monoacid ring A (BPD-MA) in 14 dogs with oral and nasal tumors. J Vet Med Sci. 2009;71(2):125–32.
23. Jacobs TM, Rosen GM. Photodynamic therapy as a treatment for esophageal squamous cell carcinoma in a dog. J Am Anim Hosp Assoc. 2000;36(3):257–61.
24. Tanabe S, Yamaguchi M, Iijima M, Nakajima S, Sakata I, Miyaki S, et al. Fluorescence detection of a new photosensitizer, PAD-S31, in tumour tissues and its use as a photodynamic treatment for skin tumours in dogs and a cat: a preliminary report. Vet J. 2004;167(3):286–93.
25. Burch S, London C, Seguin B, Rodriguez C, Wilson BC, Bisland SK. Treatment of canine osseous tumors with photodynamic therapy: a pilot study. Clin Orthop Relat Res. 2009;467(4):1028–34.
26. Lucroy M. Photodynamic therapy: potential applications for exotic animal oncology. Semin Avian Exot Pet Med. 2005;14(3):205–11.
27. Giuliano EA, MacDonald I, McCaw DL, Dougherty TJ, Klauss G, Ota J, et al. Photodynamic therapy for the treatment of periocular squamous cell carcinoma in horses: a pilot study. Vet Ophthalmol. 2008;11 Suppl 1:27–34.
28. Barnes LD, Giuliano EA, Ota J. Cellular localization of Visudyne (R) as a function of time after local injection in an in vivo model of squamous cell carcinoma: an investigation into tumor cell death. Vet Ophthalmol. 2010;13:158–65.
29. Giuliano EA, Johnson PJ, Delgado C, Pearce JW, Moore CP. Local photodynamic therapy delays recurrence of equine periocular squamous cell carcinoma compared to cryotherapy. Vet Ophthalmol. 2014;17 Suppl 1:37–45.
30. Martens A, de Moor A, Waelkens E, Merlevede W, De Witte P. In vitro and in vivo evaluation of hypericin for photodynamic therapy of equine sarcoids. Vet J. 2000;159(1):77–84.
31. Reschke C. Successful treatment of an equine sarcoid. Case report on a combined surgical and photodynamic therapy. Tierärztliche Praxis Ausgabe G Grosstiere/Nutztiere. 2012; 40(5):309–13.

32. D'Hallewin MA, Kochetkov D, Viry-Babel Y, Leroux A, Werkmeister E, Dumas D, et al. Photodynamic therapy with intratumoral administration of lipid-based mTHPC in a model of breast cancer recurrence. Lasers Surg Med. 2008;40(8):543–9.
33. Hage R, Mancilha G, Zângaro RA, Munin E, Plapler H. Photodynamic therapy (PDT) using intratumoral injection of the 5-aminolevulinic acid (5-ALA) for the treatment of eye cancer in cattle. D, editor. 2007, p. 64271C – 64271C – 7. Proc. SPIE 6427, Optical Methods for Tumor Treatment and Detection: Mechanisms and Techniques in Photodynamic Therapy XVI, 64271C.
34. Suedmeyer WK, Henry C, McCaw D, Boucher M. Attempted photodynamic therapy against patagial squamous cell carcinoma in an African rose-ringed parakeet (Psittacula krameri). J Zoo Wildl Med. 2007;38(4):597–600.
35. Rosenthal K, Duda L, Ivey E, Al E. A report of photodynamic therapy for squamous cell carcinoma in a cockatiel. In: Scientific proceedings 22nd annual association of avian veterinarians conference Orlando. 2001, p. 175–6.
36. Roberts WG, Klein MK, Loomis M, Weldy S, Berns MW. Photodynamic therapy of spontaneous cancers in felines, canines, and snakes with chloro-aluminum sulfonated phthalocyanine. J Natl Cancer Inst. 1991;83(1):18–23.
37. Ramsay E, Munson L, Lowenstine L, Al E. A retrospective study of neoplasia in a collection of captive snakes. J Zoo Wildl Med. 1996;27(1):28–34.

Chapter 11
Basic Studies in Antimicrobial PDT

Cristiane Lassálvia Nascimento, Fábio Parra Sellera, and Martha Simões Ribeiro

Abstract Microorganisms (MO) multiply rapidly and mutations that guarantee its survival forward to antimicrobials become prevalent in new populations. The inexorable rise of multidrug-resistant MO leads to an effort to search for alternative approaches that, hypothetically, MO could not easily develop resistance. Antimicrobial photodynamic therapy (PDT) is an effective alternative treatment for infected lesions in animals. The goal of the technique is to destroy a sufficient number of pathogenic MO to prevent recolonization and avoid unacceptable destruction of the host tissue. An important observation concerns the selectivity of the photosensitizer by microbial cells when compared to the host. This is because the photosensitizer (PS) uptake by host cells is slower than by MO. If the site of infection is irradiated after a short interval from the PS application (minutes), the damage to host tissue is minimized. Currently, antimicrobial PDT has proven its effectiveness against bacteria, virus, fungi, and parasites. This chapter reviews the literature regarding antimicrobial PDT especially for veterinary medicine.

C.L. Nascimento (✉)
Veterinary Unit of Santos Aquarium, Santos Aquarium, Ponta da Praia, Santos, Brazil
e-mail: cristianelassalvia@hotmail.com

F.P. Sellera
Department of Internal Medicine, School of Veterinary Medicine and Animal Science, University of São Paulo, Butantã, São Paulo, SP, Brazil
e-mail: fsellera@usp.br

M.S. Ribeiro
Center for Lasers and Applications, Nuclear and Energy Research Institute, National Commission for Nuclear Energy, São Paulo, SP, Brazil
e-mail: marthasr@usp.br

F.P. Sellera et al. (eds.), *Photodynamic Therapy in Veterinary Medicine: From Basics to Clinical Practice*, DOI 10.1007/978-3-319-45007-0_11

11.1 Introduction

One of the factors that led to disinterest for decades of alternative antimicrobial therapies such as photodynamic therapy (PDT) was the discovery of penicillin and especially the widespread use of antibiotics during World War II. Researchers imagined that the battle against microorganisms (MO) was won. However, 2 years after the commercial launch of penicillin, the emergence of resistant bacteria strains was reported [1, 2].

Bacteria multiply rapidly and mutations that guarantee its survival forward to antibiotics become prevalent in new populations in addition to other abilities as transfer of plasmids and genetic elements across species. The inappropriate prescription of antibiotics, especially for viral diseases, leads to treatment failure, and the indiscriminate use in animal production for food contributes to exacerbation of the problem by repeated selection of resistant strains. The inexorable rise of multidrug-resistant bacteria drives to an effort to search for alternative forms of treatment which, hypothetically, bacteria cannot easily develop resistance [3].

PDT has already been investigated in the treatment of skin tumors in veterinary medicine and can offer an effective alternative treatment of infected lesions in animals [4]. This technique is based on photooxidation of biological material using a photosensitizer (PS), usually a nontoxic dye introduced into the first treatment step, followed by exposure to light of suitable wavelength to be absorbed by the PS. This combination associated to oxygen results in the production of reactive oxygen species that cause damage to cellular functions leading to death [5]. When the cells to be eliminated are pathogenic MO, this procedure may be called photodynamic inactivation [3].

For the PDT to be effective as a therapeutic modality, it should be able to destroy MO in the infection site, surrounded by proteins, cells, blood, and other tissues [3]. Some studies have proven that the effectiveness of PDT becomes reduced in the presence of these components [6–9]. However, the adjustment of light parameters (power, energy, and exposure time) and correct choice of the PS have proved to be effective in different tissues: inactivation of methicillin-resistant *Staphylococcus aureus* (MRSA) in the skin [9, 10], viral inactivation of blood and components [11–15], fungal inactivation in nails [16], root canal disinfection in teeth [17, 18], and *Helicobacter mustelae* inactivation in explanted gastric mucosa of ferrets [19].

11.2 PDT and Bacteria

Literature reports that the photodynamic inactivation of multidrug-resistant bacteria is as likely as their less pathogenic equivalent [20]. Moreover, there are no reports about resistance to photodynamic inactivation of any MO tested [21].

Because the light delivery is localized, currently PDT is more suitable for local than systemic infections (such as bacteremia), applying the PS in the infected area by topical application, instillation, injection interstitial, or aerosol. The goal of the technique is to destroy pathogenic MO through an effective selectivity of the photosensitizer to prevent recolonization and avoid unacceptable destruction of the host tissue [20].

Since 1960s, it is well known that MO can be inactivated by the combination of dyes and light. In the 1990s, a fundamental difference in susceptibility to PDT between Gram-positive and Gram-negative bacteria was observed: neutral, anionic, or cationic PS efficiently inactivate Gram-positive bacteria. However, only cationic PS was able to kill Gram-negative species (or strategies to become permeable bacterial barrier to non-cationic PS). This difference in susceptibility between bacteria species is explained by their cell wall. Gram positive have a porous cytoplasmic membrane composed of a thick multilayered peptidoglycan layer and low lipid and lipoprotein content, which allows the penetration of the PS. On the other hand, Gram-negative bacteria have thin peptidoglycan layer (single-layered), periplasmic space, and a high content of lipid, lipoprotein, and lipopolysaccharides due to the outer membrane, which form an effective barrier restricting the penetration of many PS [20].

Cationic PS (i.e., molecules positively charged at physiological pH values) as the phenothiazines, methylene blue (MB) and toluidine blue (TB), have been widely used for both Gram-positive and Gram-negative bacteria and fungal cells. For Gram-negative bacteria, the mechanism of action is explained by the displacement of divalent cations (Ca 2+ and Mg 2+) from its position on the outer membrane where they act as an anchor for the lipopolysaccharide molecules negatively charged. This weakened membrane becomes permeable to the cationic PS, which gradually increases the disruption of the barrier and the absorption of the photosensitizer for each additional bond [3].

Another important observation concerns the selectivity of the PS for bacterial cells when compared to the host. This is because the PS uptake by the host cells is slower and occurs by endocytosis, while for the MO it is relatively fast. If the site of infection is irradiated after a short interval from the PS application (minutes), the damage to the host tissue is minimized [3]. Thus, some strategies are being tested to increase the permeability of the photosensitizer by Gram-negative membrane as the use of polycationic peptide polymyxin B nonapeptide (PMBN) [21] and ethylenediaminetetraacetic acid (EDTA) [22].

Many other virulence factors increase the pathogenicity such as mobility, adhesion to various tissue components, and resistance to host defense mechanisms. One characteristic that makes APDT attractive is the possibility of nonselective destruction of tissue components, particularly protein, by the action of reactive oxygen species generated by the process, making it also an anti-virulence therapy [20].

11.3 PDT and Virus

The first clinical reports of the use of photosensitizers and light against viruses have been known since 1928 which was successful in treating genital lesions of *Herpes simplex* in humans in the early 1970s [23]. However, the practice declined after reporting that the used PS (neutral red) could have carcinogenic action and did not present significant result when confronted with placebo [20]. Luckily the development of the technique was achieved with the best understanding of appropriate PS and light sources [24].

Increased concern about the risk of viral infection highlighted the importance of photodynamic inactivation of viruses, in particular the disinfection of blood products [24]. PDT has been used successfully to inactivate AIDS virus (HIV); human hepatitis A, B, and C virus (HAV, HBV, and HCV); *Cytomegalovirus*; human parvovirus; and human T-lymphotropic virus [25].

The conventional systemic or topical treatment for *Herpes simplex virus* (HSV) uses drugs that shorten the course and can suppress the virus, reducing the severity and preventing relapses. Although the incidence of viral resistance is not as serious as in the case of antibacterials, resistant forms of the virus are prevalent in immunosuppressed human patients [23] as well as in some more severe cases, and recurrent episodes occur [24]. Another aspect that needs to be considered is that the anti-HSV medication is most effective when used in prodromal phase [24]. PDT can be helpful in reducing viral titer in the vesicle phase. The main advantages appear to be the absence of side effects and drug interactions, which are especially helpful for older and immunocompromised patients [24].

Unlike HSV infections, there has been little success to find a specific antiviral therapy for human papillomatosis. Oral and genital *human papillomavirus* (HPV)-associated lesions constitute significant problems in public health [23]. Current therapeutic options include immunomodulators after laser ablation. Such approaches may require repeated sessions and could be painful. However, the use of PDT against lesions offers both antiviral and antitumor capabilities [23]. Papillomatosis has been treated by systemic and topical PDT in several anatomic locations [20].

11.4 PDT and Fungus

Dermatophytosis or fungal infections of the skin and nails are common problems affecting millions of people worldwide, especially in hot and humid climates, and are commonly treated with topical antifungal preparations. The superficial nature of fungal infections encourages the use of topical PDT as an alternative therapy [20].

The literature is rich in reporting *Candida* sp. inactivation, but hyphal and yeast forms of *C. albicans* showed different susceptibilities to PDT [26]. Some studies with *Trichophyton rubrum* both in vitro and in vivo are also reported [27, 28]. Photodynamic efficiency is closely related to light parameters, type of PS, and fungal species [3].

C. albicans is the most often isolated yeast from human patients. This yeast is responsible for opportunistic infections and constitutes a serious health risk to immunocompromised individuals. A study evaluated the efficacy of MB-mediated PDT to treat oral candidiasis in immunodeficient mice. The results indicated a MB-dependent effect and a complete eradication was achieved with a concentration of 450–500 μg/mL MB [29]. A more recent study showed that PDT was able to reduce fungal burden and inflammatory process in *C. albicans*-induced murine vaginitis.

A few studies report the PS molecular characteristics to make it effective in mediating PDT of various species of yeast and fungi; however, it is suggested that membrane damage and consequent increased permeability could be the cause of cell death after MB-mediated PDT on yeast [20].

11.5 PDT and Parasites

Several studies demonstrated the effectiveness of PDT to inactivate different parasites, but perhaps the studies focused on leishmaniasis treatment have obtained higher attention.

Leishmaniasis is a common zoonosis and it is considered one of the six most important diseases in public health. The genus *Leishmania* is a protozoan parasite transmitted by the bite of a sand fly from subfamily Phlebotominae. Cutaneous leishmaniasis is the most common form and there is no standard treatment yet [2]. The life cycle of the parasite encompasses two different forms: the promastigote, which is found in the vector and is implicated in transmission, and the amastigote, the intracellular form encountered in mammalian cells.

Different species of *Leishmania* are known to produce disease in mammals, and each species has particular biochemical and molecular characteristics that determine its taxonomy and virulence. PDT has showed to be equally effective to inactivate *L. amazonensis*, *L. braziliensis*, *L. major*, and *L. tropica* in vitro. However, different PS and distinct light sources are reported.

It is important to notice that zinc phthalocyanines (PcZn) showed different photodynamic actions for uninfected macrophages versus parasite-infected macrophage. Uninfected cells were more susceptible to PDT than infected cells [30]. Apparently, macrophage photosensitivity did not impair wound recovery. In fact, significant results were found in patients who received weekly PDT mediated by 10 % ALA preparation and red light with a wavelength of 633 nm (100 J/cm^2). A red light dose of 75 J/cm^2 performed for 12 weeks also showed good result. PDT showed to be more effective than topical paromomycin and methylbenzethonium chloride in the therapy of cutaneous leishmaniasis [3].

More recent works have showed that PDT associated with pentavalent antimonial exhibited faster wound repair [31]) and daylight-activated PDT was effective to treat cutaneous leishmaniasis [32].

Although the first PDT studies on leishmaniasis were performed in humans, some animals, including dogs and cats, are important reservoirs and may develop the disease. In this context, leishmaniasis has a special attention due to clinical importance in companion animals. Thus, veterinary medicine studies focused on the treatment of animals gain a special meaning, not only for epidemiological control of the disease but also to clinical outcomes in veterinarian patients.

11.6 PDT and Biofilms

Biofilm is characterized by a complex microbial community of cells embedded in an extracellular matrix of polysaccharide. When the cells are organized in biofilms, they have distinct phenotypic properties resulting in less MO susceptibility than in the planktonic phase [33, 34].

Studies reported that microorganisms organized in biofilms are also less susceptible to photodynamic procedure compared with planktonic phase. The lower susceptibility of microbial biofilms to PDT is probably due to heterogeneity of the biofilm, reduced growth rate of cells inside the biofilm, differences in gene expression, and limited penetration of antimicrobial agents across the extracellular matrix material [34].

One study with *Candida albicans* by Sousa et al. states that the main difference between a suspension of *Candida* and biofilm is the extracellular matrix that provides structural and biochemical support to the surrounding cells, along with a protective wall [35]. This extra protection is anionic; thus, the MB (which is cationic) concentrates within the extracellular matrix – not in or close to the *Candida* cells. This phenomenon reduces the amount of MB available near the cells, therefore decreasing the efficiency of the PDT. Since the diffusion of the MB throughout the extracellular matrix is slowed, the time of PS in contact to the yeast only allowed the MB to reach cells closer to the outer wall of the extracellular matrix, while the inner cells are well protected against the photoinactivation. Thus, these cells remain viable regardless of the irradiation time. In another study, Rossoni et al. found differences in hydrophobicity between two different serotypes of *Candida albicans*, which could explain differences in sensitivities [34].

Protocol and light-photosensitizer parameters are critical to implement PDT as a predicted treatment on clinical environment. In a biofilm this is especially important because the extracellular matrix compromises – if not completely depletes – the diffusion of the drug. Thus, despite the efficiency in generating reactive oxygen species, the effect of photosensitizers is null if the molecules are far from the cells [35].

Sousa et al. demonstrated that MB-mediated PDT reduced *Candida* biofilm by approximately 96 % after 10 min of irradiation [35]. Rossoni et al. reported that MB-mediated PDT with red laser radiation exerted a fungicidal effect on biofilms of *C. albicans* serotypes A and B and that serotype B was more sensitive than serotype A [34]. In addition, serotype B was also sensitive to laser application alone. Farias et al. showed that PDT inhibited insoluble extracellular polysaccharides in biofilm of *Streptococcus mutans* through loss of bacterial viability [33].

All those studies indicate PDT as a promising tool against MO biofilms, since adequate parameters are attended.

11.7 Basic Studies in Veterinary Medicine

The first in vitro studies for MO inactivation were carried out between the 1990s and the early 2000s, when researchers investigated the potential of PDT to inactivate pathogens from blood, semen, and embryos.

Carpenter and Kraus investigated the effect of hypericin associated with fluorescent bulbs to control *Equine infectious anemia virus* in vitro [36]. The authors demonstrated that the action of hypericin was dependent on light. Short periods of photosensitivity resulted in a partial loss of reverse transcriptase activity and complete inhibition of viral infectivity. Authors concluded that the photodynamic effect of hypericin interferes in more than one stage in the virus replication cycle.

Bielanski et al. investigated hematoporphyrin (HP), hematoporphyrin derivative (HPD), and thiopyronine (TP) as photosensitizers activated by a He-Ne red laser (HP, HPD), white microscope light (HP, HPD), or yellow-green microscope light (TP) [37].

Experiments were conducted with appropriate controls to determine the effect of the PS to inactivate *bovine herpesvirus 1* (BHV-1) and *bovine viral diarrhea virus* (BVDV), to disinfect day-7 zona pellucida-intact bovine embryos that had been exposed to BHV-1 or BVDV, and on the in vitro development of embryos. Exposure to HP, HPD, and TP followed by light irradiation inactivated BHV-1 and BVDV. Embryos exposed to BHV-1 were disinfected by HP or HPD in combination with the He-Ne laser or by HP or HPD in combination with white light. Embryos exposed to BVDV were disinfected by HPD followed by He-Ne or white light irradiation. Exposure of embryos to light alone or to light and HP or HPD had no detrimental effect on their in vitro development. However, exposure of embryos to TP followed by irradiation caused embryonic degeneration. Authors concluded that exposure of embryos to HP/HPD followed by irradiation with He-Ne laser or white light is a quick, simple, and effective method for disinfection of embryos carrying BHV-1 or BVDV.

Following that study, Eaglesome et al. investigated the association of photosensitizers (HP, HPD, and TP) and light sources (red laser, white light, and yellow-green light) for inactivation of pathogenic MO in culture medium and in bovine semen [38]. Bovine microbial pathogens were suspended in tissue culture medium and/or PBS and also added to bovine semen. Then, they were exposed to the PS followed by irradiation. HP, HPD, and TP were effective against BHV-1, BVDV, *Mycoplasma bovigenitalium*, *Mycoplasma canadense*, and *Ureaplasma diversum* in culture media. In addition, TP was effective against *Leptospira pomona*. Similar treatments were not effective against *Leptospira hardjo*, *Mycoplasma bovis*, or *Campylobacter fetus* subsp, *venerealis*. When MO were added to bovine semen, only BHV-1 was controlled by the PS at concentrations that did not appear harmful to sperm cells.

In another study, Washburn et al. observed the effects of illumination on the virucidal properties of MB followed by exposure to white fluorescent lamps to inactivate the *Caprine arthritis-encephalitis virus* in translucent medium as well as colostrum [39]. Although the viral model used was BVDV, the authors concluded than 60 min of treatment significantly reduced virus levels in samples containing 0.01 μM MB compared to treatment without light. Moreover, the immunoglobulin concentrations were not adversely affected. The authors concluded that PDT seems feasible to treat the colostrum of goats.

Another application of PDT in veterinary attempted to eliminate the *Feline leukemia virus* (FeLV) with benzoporphyrin derivative monoacid ring A (BPD-MA) and red light. Although this in vitro study has used blood from experimentally infected cats, there was selective photoinactivation of infected cells and disappearance of the virus. In that study, the authors demonstrated that T cells infected in culture were slightly more sensitive to PDT than uninfected cells, indicating new possibilities to inactivate retroviruses [40].

An interesting study that deserves special attention by veterinary dermatology was conducted by Wardlaw et al. [41]. Those authors evaluated the effects of PDT in vitro against *Staphylococcus intermedius, Pseudomonas aeruginosa, Streptococcus canis*, and *Escherichia coli*, bacteria commonly found in skin infections. ALA diluted in 0.9% saline was placed in bacterial cultures to induce protoporphyrin IX (PpIX) production, and MO were irradiated with incoherent light source (wavelength of 635 nm with an irradiance of 15 mW/cm^2 and different dosages depending on exposure time). The results showed that control groups (no light and no ALA, light alone, ALA alone) did not significantly alter bacterial survival at 1, 2, 3, 4, or 6 h of exposure. Compared with the controls, there was a significant decrease in bacterial survival after PDT for all MO excluding *E. coli*.

Recently, Pires et al. demonstrated the PDT effectiveness in vitro and in vivo for *Pythium insidiosum* inactivation [42]. For in vitro studies, two photosensitizers were evaluated: Photogem® and chlorine. Amphotericin B was also evaluated, and the control group was treated with sterile saline solution. For in vivo studies, six rabbits were inoculated with 20,000 zoospores of *P. insidiosum*, and an area of 1 cm^2 was treated using the same PSs. The irradiation was performed using a laser emitting at 660 nm and a radiant exposure of 200 J/cm^2. Rabbits were clinically evaluated daily and histopathological analysis was performed 72 h after PDT. For in vitro assays, PDT showed inhibition rates from 60 to 100 %. Photogem® and chlorine showed better results in comparison to amphotericin B. For the in vivo assays, histological analysis of lesions showed a lack of infection up to 1 cm in depth.

That study incited the development of new studies involving *P. insidiosum*. Pires et al. investigated in vitro PDT response on the growing of *P. insidiosum* culture using other PSs: MB, Photogem®, and Photodithazine® [43]. The photosensitizer distribution in cell structures and the PDT response for incubation times of 30, 60, and 120 min were evaluated. MB did not spread through the cell and, consequently, there was no PDT response. Photogem® showed heterogenous distribution in the hyphal structure with small concentration inside the cells, and dead and live cells were observed in the treated culture. After 48 h, hyphal regrowth was observed. Photodithazine® showed more homogenous distribution inside the cell with specific intracellular localization dependent on the incubation time. Photodithazine® first accumulates in intracellular vacuoles, and at incubation times of 1 h, it is located at all cell membrane. An inhibition over 98 % of the growing rates was achieved with Photodithazine-mediated PDT. The authors concluded that the PSs that cross more efficiently the *P. insidiosum* membranes are able to cause extensive damage to the organism under irradiation and therefore are the best options for clinical treatment. Despite the good results obtained against *P. insidiosum*, no clinical studies of PDT in equines were described until now.

Nascimento and collaborators tested PDT using a red light-emitting diode (LED) and MB against four resistant MO isolated from bumblefoot: *Staphylococcus aureus*, *Escherichia coli*, *Pseudomonas aeruginosa*, and *Proteus mirabilis* [44]. Their findings showed effective bacterial reduction for *P. aeruginosa, P. mirabilis*, and *S. aureus* depending on irradiation time. Under tested parameters in that study, PDT was not able to efficiently inactivate *E. coli*.

In addition, the use of PDT should be highlighted to control bovine mastitis, given the importance of this disease in the production of milk and concern about the use of antibiotics. We investigated the ability of PDT in vitro to inactivate pathogens associated with infectious bovine mastitis [45]. Antibiotic-resistant strains, isolated from bovine mastitis as *Staphylococcus aureus*, *Streptococcus agalactiae*, *Streptococcus dysgalactiae*, *Corynebacterium bovis*, and the algae *Prototheca zopfii*, were tested. Nine experimental groups were evaluated: control without any treatment, light only (a red LED ($\lambda = 662$ nm) for 180 s), PS only (50 μM MB for 5 min), and PDT for 5, 10, 30, 60, 120, and 180 s. *S. dysgalactiae*, *S. aureus*, and *C. bovis* were inactivated after 30 s of irradiation, whereas *S. agalactiae* was inactivated after 120 s and *P. zopfii* at 180 s of irradiation. Although the results are limited to laboratory experiments, PDT deserves further investigation for this purpose, especially in chronic infections or unresponsive, as in the case of *P. zopfii*. A proper tool of light delivery should be developed to illuminate all structures involved, since the MO might be distributed throughout the entire mammary gland (Fig. 11.1).

Despite the lack of standardization regarding light parameters, PS and MO, the basic studies reported in this chapter encourage the use of PDT in clinical practice. As you will see in the next chapter, PDT could be an adjuvant or alternative approach to infectious diseases of difficult treatment.

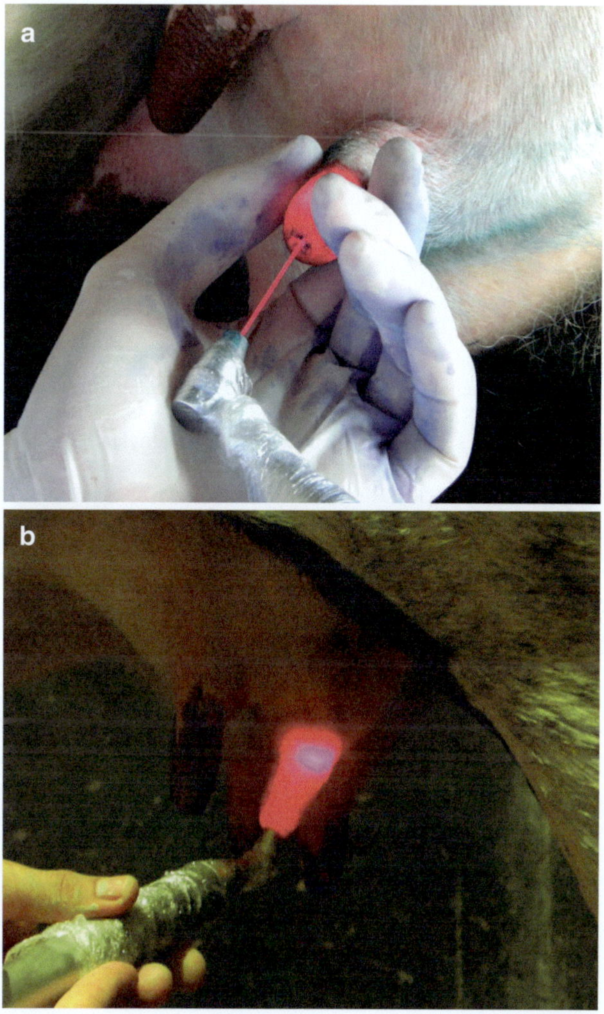

Fig. 11.1 Bovine teat under irradiation. (**a**) Use of an optical fiber in the teat canal to access the mammary gland and (**b**) PDT procedure. Note the teat completely illuminated but not the entire mammary gland

References

1. Yoshikawa TT. Antimicrobial resistance and aging: beginning of the end of the antibiotic era? J Am Geriatr Soc. 2002;50(7 Suppl):S226–9.
2. Bisht R, Katiyar A, Singh R, Mittal P. Antibiotic resistance – a global issue of concern. Asian J Pharm Clin Res. 2009;2(2):34–9.
3. Dai T, Huang Y-Y, Hamblin MR. Photodynamic therapy for localized infections – state of the art. Photodiagnosis Photodyn Ther. 2009;6(3–4):170–88.
4. Vinck E, Cagme B, Vinck H, Cambier D. Is photodynamic therapy an appropriate treatment of feline superficial squamous cell carcinomas? Two case studies in small animal practice. In: Longo L Hofstetter AG, Pascu M-L, Waidelich WRA, editors. Proceedings of SPIE. Bellingham; 2003, p. 39–45.
5. Henderson BW, Dougherty TJ. How does photodynamic therapy work? Photochem Photobiol. 1992;55(1):145–57.
6. Nitzan Y, Balzam-Sudakevitz A, Ashkenazi H. Eradication of acinetobacter baumannii by photosensitized agents in vitro. J Photochem Photobiol B Biol. 1998;42(3):211–8.
7. Wilson M, Pratten J. Lethal photosensitisation of Staphylococcus aureus in vitro: effect of growth phase, serum, and pre-irradiation time. Lasers Surg Med. 1995;16(3):272–6.
8. Lambrechts SAG, Aalders MCG, Verbraak FD, Lagerberg JWM, Dankert JB, Schuitmaker JJ. Effect of albumin on the photodynamic inactivation of microorganisms by a cationic porphyrin. J Photochem Photobiol B Biol. 2005;79(1):51–7.
9. Street CN, Pedigo L, Gibbs A, Loebel NG. Antimicrobial Photodynamic Therapy for the Decolonization of Methicillin-Resistant Staphylococcus aureus from the Anterior Nares. In: Kessel DH, editor. Photodynamic therapy: Back to the Future. Seattle: Proc. SPIE 7380; 2009. p. 73803B – 73803B – 16.
10. Maisch T, Bosl C, Szeimies R-M, Love B, Abels C. Determination of the antibacterial efficacy of a new porphyrin-based photosensitizer against MRSA ex vivo. Photochem Photobiol Sci. 2007;6(5):545–51.
11. Mohr H, Bachmann B, Klein-Struckmeier A, Lambrecht B. Virus inactivation of blood products by phenothiazine dyes and light. Photochem Photobiol. 1997;65(3):441–5.
12. Mohr H, Knüver-Hopf J, Gravemann U, Redecker-Klein A, Müller TH. West Nile virus in plasma is highly sensitive to methylene blue-light treatment. Transfusion. 2004;44(6):886–90.
13. Mohr H, Redecker-Klein A. Inactivation of pathogens in platelet concentrates by using a two-step procedure. Vox Sang. 2003;84(2):96–104.
14. Ben-Hur E, Geacintov NE, Studamire B, Kenney ME, Horowitz B. The effect of irradiance on virus sterilization and photodynamic damage in red blood cells sensitized by phthalocyanines. Photochem Photobiol. 1995;61(2):190–5.
15. Ben-Hur E, Barshtein G, Chen S, Yedgar S. Photodynamic treatment of red blood cell concentrates for virus inactivation enhances red blood cell aggregation: protection with antioxidants. Photochem Photobiol. 1997;66(4):509–12.
16. Smijs TGM, Bouwstra JA, Schuitmaker HJ, Talebi M, Pavel S. A novel ex vivo skin model to study the susceptibility of the dermatophyte Trichophyton rubrum to photodynamic treatment in different growth phases. J Antimicrob Chemother. 2007;59(3):433–40.
17. Fonseca MB, Júnior POT, Pallota RC, Filho HF, Denardin OVP, Rapoport A, et al. Photodynamic therapy for root canals infected with Enterococcus faecalis. Photomed Laser Surg. 2008;26(3):209–13.
18. Foschi F, Fontana CR, Ruggiero K, Riahi R, Vera A, Doukas AG, et al. Photodynamic inactivation of Enterococcus faecalis in dental root canals in vitro. Lasers Surg Med. 2007;39(10):782–7.
19. Millson CE, Wilson M, MacRobert AJ, Bown SG. *Ex-vivo* treatment of gastric Helicobacter infection by photodynamic therapy. J Photochem Photobiol B Biol. 1996;32(1–2):59–65.

20. Hamblin MR, Hasan T. Photodynamic therapy: a new antimicrobial approach to infectious disease? Photochem Photobiol Sci. 2004;3(5):436–50.
21. Nitzan Y, Gutterman M, Malik Z, Ehrenberg B. Inactivation of gram-negative bacteria by photosensitized porphyrins. Photochem Photobiol. 1992;55(1):89–96.
22. Bertoloni G, Rossi F, Valduga G, Jori G, van Lier J. Photosensitizing activity of water- and lipid-soluble phthalocyanines on Escherichia coli. FEMS Microbiol Lett. 1990;59(1–2):149–55.
23. Wainwright M. Local treatment of viral disease using photodynamic therapy. Int J Antimicrob Agents. 2003;21(6):510–20.
24. de Paula EC, Aranha ACC, Simões A, Bello-Silva MS, Ramalho KM, Esteves-Oliveira M, et al. Laser treatment of recurrent herpes labialis: a literature review. Lasers Med Sci. 2014;29(4):1517–29.
25. Costa L, Faustino MAF, Neves MGPMS, Cunha A, Almeida A. Photodynamic inactivation of Mammalian viruses and bacteriophages. Viruses. 2012;4(7):1034–74.
26. Jackson Z, Meghji S, MacRobert A, Henderson B, Wilson M. Killing of the yeast and hyphal forms of candida albicans using a light-activated antimicrobial agent. Lasers Med Sci. 1999;14(2):150–7.
27. Baltazar L de M, Soares BM, Carneiro HCS, Avila TV, Gouveia LF, Souza DG, et al. Photodynamic inhibition of Trichophyton rubrum: in vitro activity and the role of oxidative and nitrosative bursts in fungal death. J Antimicrob Chemother. 2013;68(2):354–61.
28. Baltazar LM, Werneck SMC, Carneiro HCS, Gouveia LF, de Paula TP, Byrro RMD, et al. Photodynamic therapy efficiently controls dermatophytosis caused by Trichophyton rubrum in a murine model. Br J Dermatol. 2015;172(3):801–4.
29. Teichert MC, Jones JW, Usacheva MN, Biel MA. Treatment of oral candidiasis with methylene blue-mediated photodynamic therapy in an immunodeficient murine model. Oral Surg Oral Med Oral Pathol Oral Radiol Endod. 2002;93(2):155–60.
30. de Silva O EP, Mittmann J, Ferreira VTP, Cardoso MAG, Beltrame M. Photodynamic effects of zinc phthalocyanines on intracellular amastigotes of Leishmania amazonensis and Leishmania braziliensis. Lasers Med Sci. 2015;30(1):347–54.
31. Song D, Lindoso JAL, Oyafuso LK, Kanashiro EHY, Cardoso JL, Uchoa AF, et al. Photodynamic therapy using methylene blue to treat cutaneous leishmaniasis. Photomed Laser Surg. 2011;29(10):711–5.
32. Enk CD, Nasereddin A, Alper R, Dan-Goor M, Jaffe CL, Wulf HC. Cutaneous leishmaniasis responds to daylight-activated photodynamic therapy: proof of concept for a novel self-administered therapeutic modality. Br J Dermatol. 2015;172(5):1364–70.
33. de Sousa Farias SS, Nemezio MA, Corona SAM, Aires CP, Borsatto MC. Effects of low-level laser therapy combined with toluidine blue on polysaccharides and biofilm of Streptococcus mutans. Lasers Med Sci. 2016;31(5):1011–6.
34. Rossoni RD, Barbosa JO, de Oliveira FE, de Oliveira LD, Jorge AOC, Junqueira JC. Biofilms of Candida albicans serotypes A and B differ in their sensitivity to photodynamic therapy. Lasers Med Sci. 2014;29(5):1679–84.
35. Sousa AS, Prates RA, de Santi MESO, Lopes RG, Bussadori SK, Ferreira LR, et al. Photodynamic inactivation of Candida albicans biofilm: influence of the radiant energy and photosensitizer charge. Photodiagnosis Photodyn Ther. 2016;14:111–4.
36. Carpenter S, Kraus GA. Photosensitization is required for inactivation of equine infectious anemia virus by hypericin. Photochem Photobiol. 1991;53(2):169–74.
37. Bielanski A, Dubuc C, Hare WCDCD, Myers DJJ, Eaglesome MDD. Inactivation of bovine herpesvirus-1 and bovine viral diarrhea virus in association with preimplantation bovine embryos using photosensitive agents. Theriogenology. 1992;38(4):633–44.
38. Eaglesome MD, Bielanski A, Hare WCD, Ruhnke HL. Studies on inactivation of pathogenic microorganisms in culture media and in bovine semen by photosensitive agents. Vet Microbiol. 1994;38(3):277–84.
39. Washburn K, Streeter R, Saliki J, Lehenbauer T, Prado M. Photodynamic inactivation of an RNA enveloped virus in goat colostrum. Small Rumin Res. 2001;42(1):31–7.

40. North J, Freeman S, Overbaugh J, Levy J, Lansman R. Photodynamic inactivation of retrovirus by benzoporphyrin derivative: a feline leukemia virus model. Transfusion. 1992;32(2):121–8.
41. Wardlaw JL, Sullivan TJ, Lux CN, Austin FW. Photodynamic therapy against common bacteria causing wound and skin infections. Vet J. 2012;192(3):374–7.
42. Pires L, Bosco S de MG, da Silva NF, Kurachi C. Photodynamic therapy for pythiosis. Vet Dermatol. 2013;24(1):130–6.e30.
43. Pires L, Bosco S de MG, Baptista MS, Kurachi C. Photodynamic therapy in Pythium insidiosum – an in vitro study of the correlation of sensitizer localization and cell death. PLoS One. 2014;9(1):e85431.
44. Nascimento CL, Ribeiro MS, Sellera FP, Dutra GH, Simões A, Teixeira CR. Comparative study between photodynamic and antibiotic therapies for treatment of footpad dermatitis (bumblefoot) in Magellanic penguins (Spheniscus magellanicus). Photodiagnosis Photodyn Ther. 2015;12(1):36–44.
45. Sellera FP, Sabino CP, Ribeiro MS, Gargano RG, Benites NR, Melville PA, et al. In vitro photoinactivation of bovine mastitis related pathogens. Photodiagnosis Photodyn Ther. 2016;13:276–81.

Chapter 12
Clinical Applications of Antimicrobial PDT

Fábio Parra Sellera, Cristiane Lassálvia Nascimento, and Martha Simões Ribeiro

Abstract Antimicrobial photodynamic therapy (PDT) is a noninvasive, painless, and safe procedure that can be used on a wide range of diseases in clinical practice. Dermatology for dogs and cats is emerging as one of the main areas of interest for the application of this therapeutic approach; however, despite being a promising technique to treat various dermatological diseases, researchers and clinicians have made little use of this powerful tool. The farm animals' medicine has evolved over the past few years, and the growing demand of consumers looking for absence of residues in animal products becomes essential in the search for alternatives to conventional treatments. Although the bulk of researches on PDT in veterinary medicine are to treat domestic animals, there is a great appeal for its clinical application in exotic, zoo, and wildlife medicine. In this chapter, we describe potential applications of PDT in veterinary clinical practice.

12.1 Introduction

As mentioned in previous chapters, there are many reasons to use photodynamic therapy (PDT) on veterinary clinical practice, and there is a vast field to be explored. The main argument still is to avoid unnecessary use of antibiotics for animal protection and, most of all, for humans. In fact, to prevent residues in animal products is a

F.P. Sellera (✉)
Department of Internal Medicine, School of Veterinary Medicine, and Animal Science, University of São Paulo, Av. Prof. Orlando Marquês Paiva, 87, Butanta 13635-900, SP, Brazil
e-mail: fsellera@usp.br

C.L. Nascimento
Veterinary Unit of Santos Aquarium, Santos Aquarium, Av. Bartolomeu de Gusmão S/N, Ponta da Praia, Santos 11035-500, SP, Brazil
e-mail: cristianelassalvia@hotmail.com

M.S. Ribeiro
Center for Lasers and Applications, Nuclear, and Energy Research Institute, National Commission for Nuclear Energy, Av. Lineu Prestes 2242, Cidade Universitária, Sao Paulo 05508-000, SP, Brazil
e-mail: marthasr@usp.br

F.P. Sellera et al. (eds.), *Photodynamic Therapy in Veterinary Medicine: From Basics to Clinical Practice*, DOI 10.1007/978-3-319-45007-0_12

growing demand from consumers, but we also should think about other sources of resistant microorganisms (MO) as the intimate contact with companion animals or exotic pets.

At the same way, veterinarians working in zoos or rehabilitation centers of wild animals must be mindful not to introduce resistant pathogens among immunologically not challenged wildlife. Antimicrobial PDT is a noninvasive, painless, and safe approach that can be used on a wide range of diseases in clinical practice.

12.2 Companion Animals

Dermatology for dogs and cats is emerging as one of the main areas of interest for application of antimicrobial PDT. However, despite being a promising technique to treat various dermatological diseases, researchers and clinicians have made little use of this powerful approach.

A study by Fabris et al. [1] investigated the effects of PDT on pyodermitis in dogs. Dogs that had lesions no longer susceptible to antibiotic-based therapies were included in the study. In all cases, the photosensitizer (PS) was topically applied on the lesions using a porphyrin-based gel formulation. After 30–60 min, the lesion was irradiated at an irradiance of 180 mW/cm^2 for up to 30 min, depending on the lesion severity. The results obtained showed that there was no significant diffusion of the PS outside where the lesion took place. In addition, free-PS areas that received irradiation were not damaged. The response of lesions to PDT concomitantly occurred with the wound healing and microbial reduction. Unfortunately, the authors described only a few details about the photosensitizer and light parameters, making difficult the repeatability of the experiment.

Another interesting approach for PDT is the treatment of infected wounds using methylene blue (MB), due to low cytotoxicity of the PS to host cells and broad action spectrum against several pathogenic strains. Figure 12.1 shows a dog that was hit and presented necrotic tissue and secondary infection. We used 0.01 % MB applied topically, pre-irradiation time (PIT) of 5 min, red laser irradiation at $\lambda = 660$ nm, $P = 100$ mW, $E = 8$ J/point, and $\Delta t = 80$ s/point, throughout the lesion once a week. Observe that after 2 weeks, the wound is re-epithelizing and with granulation tissue formation (Fig. 12.1).

Figure 12.2 illustrates another case of PDT on infected traumatic wound of unknown cause. PDT protocol was maintained the same as abovementioned, but a single PDT was performed. The absence of necrosis and granulation tissue formation can be noticed.

PDT is well documented for the treatment of dental diseases, including clinical trials. Oral infections described in humans are commonly seen in animals. Although few studies of PDT for veterinary dentistry have been described, there is an interesting prospect for clinical application.

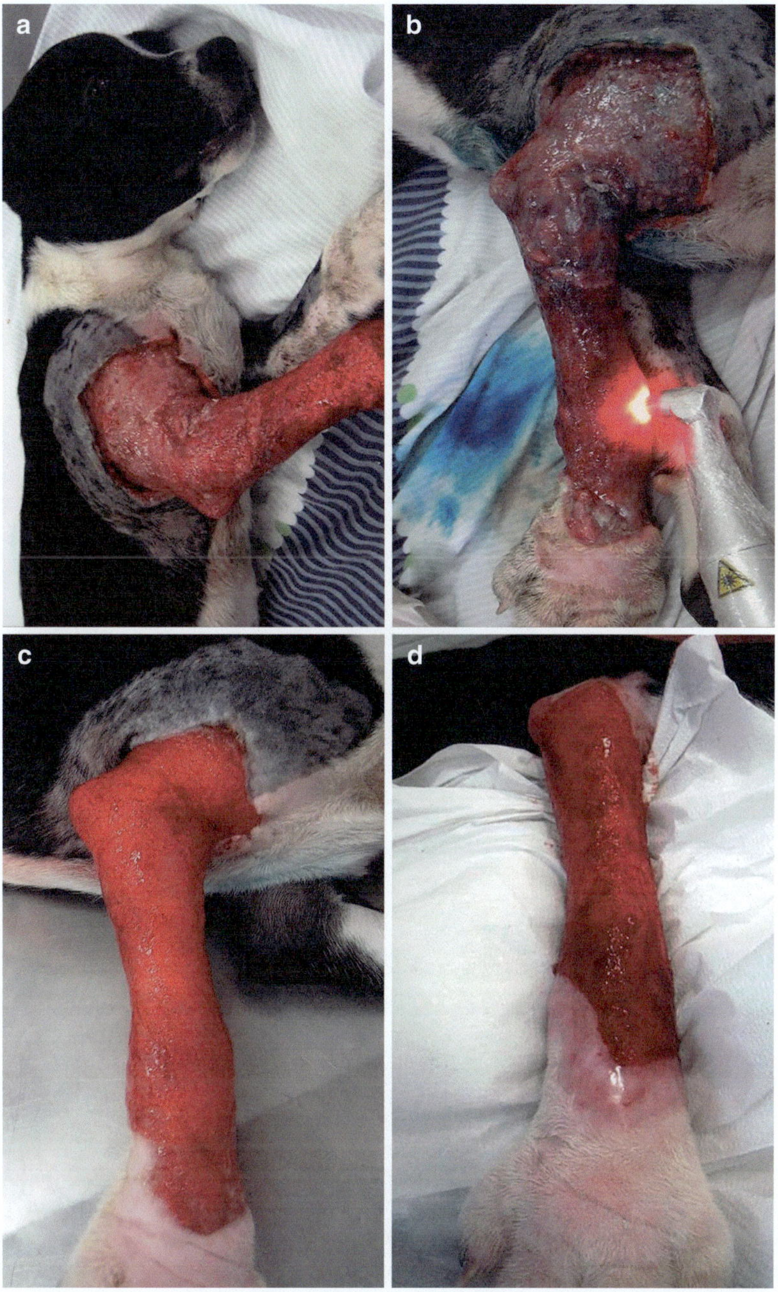

Fig. 12.1 Evolution of run-over case in a puppy treated by PDT. (**a**) Extensive wound, with the presence of necrosis and secondary infection. (**b**) PDT procedure. (**c**) Aspect of the lesion after 1 week. (**d**) Aspect of the lesion after 2 weeks. Note the absence of infection after 1 week and good reepithelialization and granulation tissue formation after 2 weeks

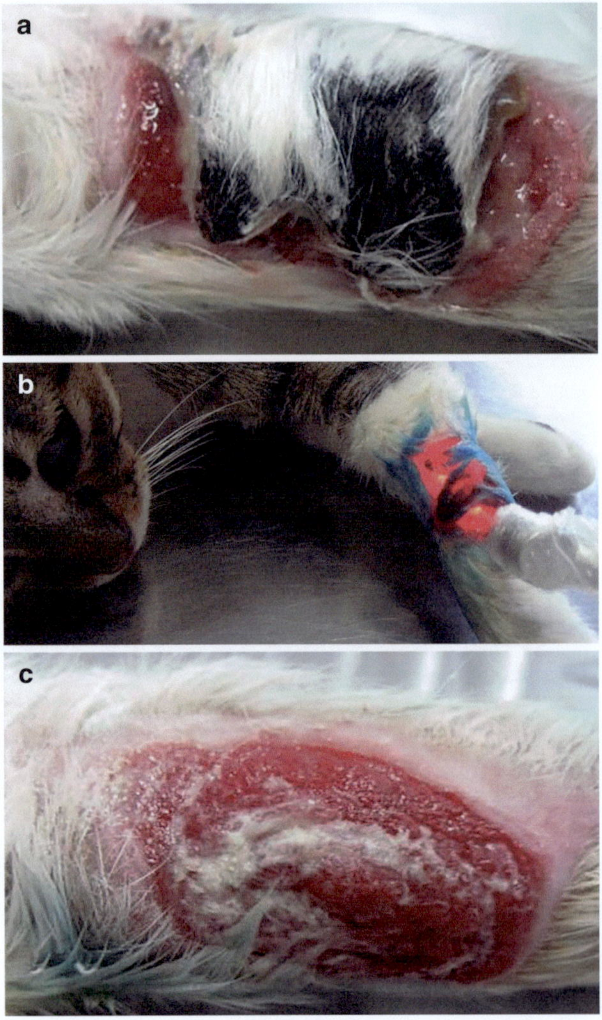

Fig. 12.2 Image of infected wound in a cat treated by PDT (**a–c**). (**a**) Initial aspect of the lesion, presenting exudate, necrosed skin flap, and inflammation signs. (**b**) PDT procedure. (**c**) Aspect of the lesion after 3 days. The necrosed skin flap was spontaneously detached from tissue beside inflammation reduction and granulation tissue formation

Hayek et al. [2] compared the effects of PDT and conventional technique on microbial reduction in ligature-induced peri-implantitis in dogs. For this study, 18 third premolars from nine Labrador retriever dogs were extracted and the implants were submerged. After osseointegration, peri-implantitis was induced. Four months later, ligature was removed and natural bacterial plaque was allowed to form for further 4 months. In the conventional group, mucoperiosteal flaps were used for scaling the implant surface and chlorhexidine irrigation. In the PDT group, only mucoperiosteal scaling was carried out before PDT. Inside the peri-implant pocket, the paste-based azulene PS was placed, and then a red laser ($\lambda=660$ nm, E=7.2 J for 3 min) was used. The results showed that *Prevotella* spp., *Fusobacterium* spp., and *Streptococcus beta-haemolyticus* were significantly reduced for both groups. The authors concluded that PDT is a noninvasive approach that could be used to reduce MO in peri-implantitis (Fig. 12.3).

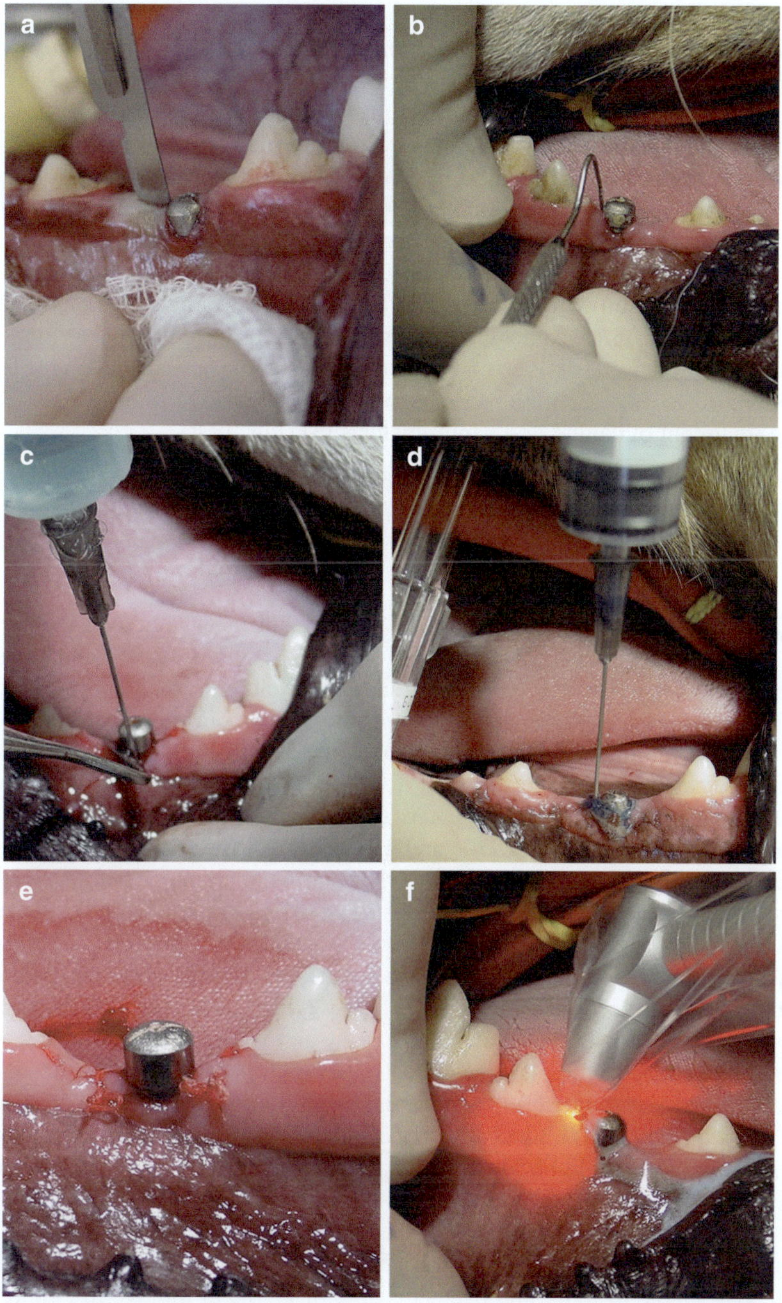

Fig. 12.3 Conventional treatment (**a–c**) and PDT (**d–f**) for induced peri-implantitis in dogs. (**a**) Elevation of mucoperiosteal flaps using a scalpel. (**b**) Irrigation with 0.12 % chlorhexidine solution after removal of peri-inflammatory granulation tissue. (**c**) Peri-implant tissue after the suture. (**d**) Mechanical implant surface debridement using a curette. (**e**) Azulene (AZ) photosensitizer (0.01 % w/w) was applied with a thin needle and left in site for 5 min. (**f**) Transmucosal illumination using a red laser

Our group also investigated a clinical protocol using PDT for the treatment of canine oral papillomatosis. It is a common disease in small animals caused by *Papillomavirus*, resulting in tumorous lesions in the oral cavity of dogs. Different therapeutic methods have been reported, but treatment remains an open question.

We treated a mixed breed adult canine with multiple verrucous oral lesions compatible with canine oral papillomatosis. Major lesions in the buccal mucosa, excluding the tongue lesions, were inoculated with an aqueous MB solution (300 μM) under general anesthesia, and after 5 min, the lesions were irradiated with a red diode laser ($\lambda = 660$ nm) for 3 min (40 mW, 0.4 J per point, 10 J/cm^2). After 15 days, a second PDT session was performed. In this time, the irradiated lesions were significantly smaller and we included the tongue lesions, which received the same protocol as the buccal lesions. A complete reduction in the buccal and tongue lesions was observed 15 days after the second application. Thus, PDT may be a new option for the treatment of canine oral papillomatosis [3] (Fig. 12.4).

12.3 Farm Animals

The farm animals' medicine, not unlike other areas, has evolved over the past few years. New technologies have been used for diagnosis and treatment of diseases in order to increase performance of these animals, for either leisure, sport, or production. Diseases of microbiological origin are among the main concerns. Many pathogens have zoonotic potential resulting in public health problems. Moreover, the growing demand of consumers looking for absence of residues in animal products becomes essential in the search for alternatives to conventional treatments.

Some PDT studies in the 1970s and 1980s against different insects, including house flies [4], motivated one of the first veterinary antimicrobial PDT studies known, which was conducted by Hawkins et al. [5]. These authors investigated the effect of erythrosin B and visible light to inactivate third stage larvae (L3) of gastrointestinal nematodes in naturally infected bovines to control larvae on pasture and prevent/reduce reinfection. In this study, cattle were treated orally with erythrosin B at dosages of 30 and 40 mg/kg/day for as many as 17 days and fecal samples from treated and untreated animals were collected per rectum daily and parasite eggs per gram of feces were counted for each animal using flotation procedure. The cultures were stored at dark until L3 had developed and, after that, were exposed to sunlight or artificial fluorescent light. The authors concluded that both sunlight and artificial fluorescent light were toxic to larvae after erythrosin B administration [5].

In a subsequent study, Hawkins et al. [6] developed another clinical assay to determine the oral dose of erythrosin B (20, 40, and 60 mg/kg/day), exposure time of infective larvae to light, and exposure time of larvae to the dye in culture. The results showed that both 60 and 40 mg/kg/day treatments produced a significant effect after 2 h of light exposure. The 20 mg/kg/day caused a significant effect after 4 h in the 30-day culture group and after 5 h in all culture groups. Furthermore, no-swimming larvae reached 97 % after 6 h of exposure to fluorescent light, approximately one-tenth of direct sunlight intensity. The authors concluded that toxic effects were strictly related to erythrosin B dosage and time of light exposure.

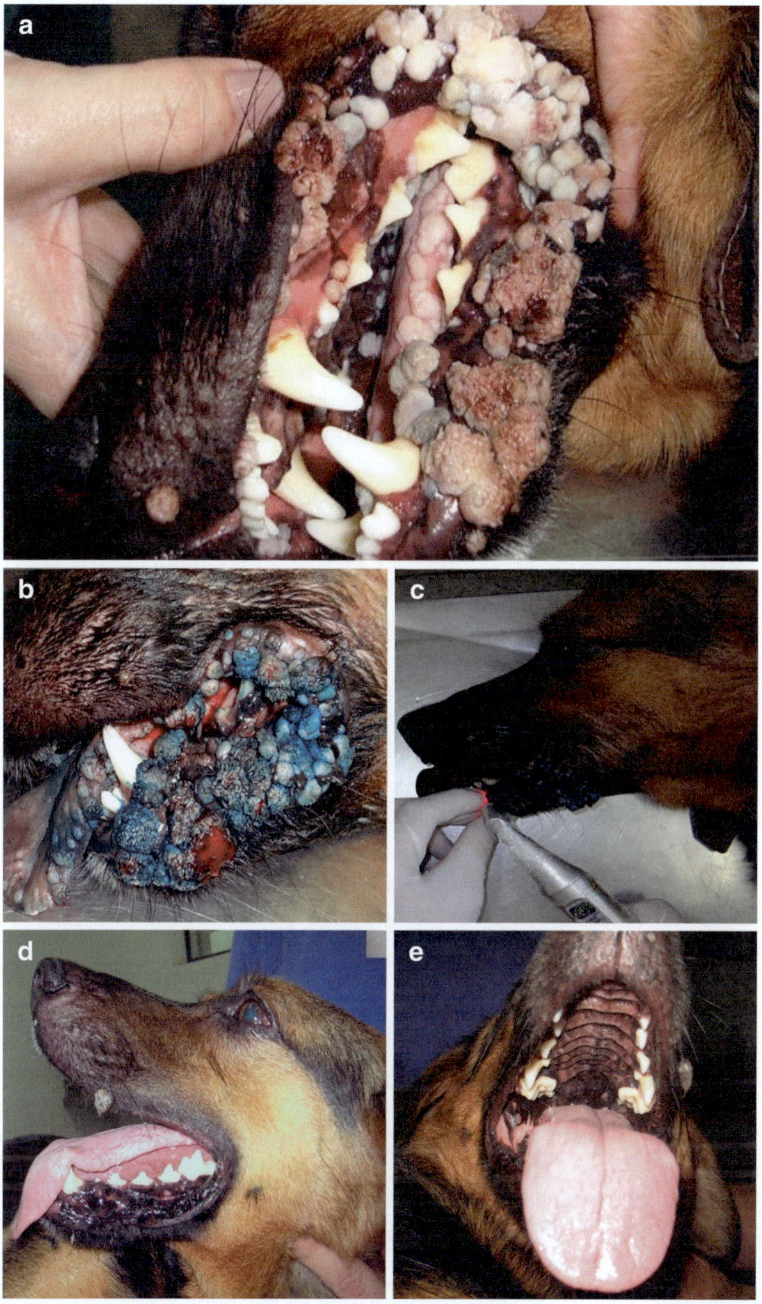

Fig. 12.4 Evolution of canine oral papillomatosis case treated by PDT. (**a**) A mixed breed adult dog presenting multiple verrucous lesions in the tongue and oral cavity. (**b**) Methylene blue was intralesionally applied after general anesthesia. (**c**) After 5 min, the lesions were irradiated with a red laser perpendicular to the lesion. (**d, e**) Complete reduction of buccal and tongue lesions after 30 days (two PDT treatments) (Copyright image International Journal of Science Commerce and Humanities)

Six years later, Healey et al. [7] made three experiments to investigate the phototoxicity of erythrosin B: (1) against L3 gastrointestinal nematodes in naturally infected ovine, (2) its persistence after oral administration, and (3) when added directly to feces.

Oral administration of 0, 40, 60, and 80 mg/kg/day for 10 consecutive days to the larvae, exposed to fluorescent light for 6 h, resulted in significant mortality, with a dose-dependent response, i.e., increased doses promoted higher mortality rates, with 16 %, 46 %, 55 %, and 62 %, respectively. The length of time that erythrosin B phototoxicity persisted was evaluated during 10 days of oral administration (80 mg/kg/day) and after 21 consecutive days of its interruption. Feces were collected and irradiated by the same light bulb and time. The L3 killing rates were significantly higher during the oral administration period, to a maximum of 76 % by day 10, but no effect was observed within 2 days after its discontinuation [7].

Feces with nematode ova that had directly received different concentrations of erythrosin B (125, 250, 500, 750, 1000, and 3000 mg/dye/kg feces) and exposed to fluorescent light for 6 h showed a dose-response curve, with higher mortalities based on the higher concentrations of dye. The lowest significant dosage was 250 mg/dye/kg feces, with 33 % of mortality rate to a maximum of 80 % at 3000 mg/dye/kg feces. The authors also emphasized that phototoxicity would be enhanced by using sunlight instead of a fluorescent light since it has only 10 % of sunlight intensity [7].

PDT clinical assays to treat farm animal's infections had its cycle interrupted in the 1990s and just returned in this last decade, over two decades after Healey's study [7].

In a study developed by our team, six male Holstein calves, aged between 4 and 5 months, were submitted to the technique of dehorning by cauterization. At the end of the procedure, all animals received a single application of PDT in the lesion of the right side, while lesions in the left side were daily treated with topic zinc oxide ointment (20 %). PDT was made with the MB (aqueous solution, 60 μM) topically applied on the lesions. After a pre-irradiation time (PIT) of 5 min, lesions were irradiated with a diode 660 nm red laser with an energy density equal to 180 J/cm^2 per point distributed in five points in each lesion. Fifteen days after dehorning, the lesions treated with PDT were cured although the ones treated daily with ointment were not completely healed and still showed signs of local inflammation [8] (Fig. 12.5).

Likewise, skin infections caused by multiresistant bacteria are susceptible to PDT treatment. We used PDT (MB and 660 nm red laser) for abscess in the left forelimb of one sheep infected by multidrug-resistant *Streptococcus* spp. previously treated with antibiotics. MB (0.01 %) was applied topically for 5 min, and the lesion was irradiated using a red diode laser ($\lambda = 660$ nm, 100 mW), coupled to an optical fiber during 180 s. The total energy used was 0.4 J per point and the energy density was 10 J/cm^2 per point distributed in five equidistant points of 0.02 cm^2. After 1 day, complete healing was achieved with no recurrence observed during the 3-month follow-up period. These results indicate PDT as a therapeutic option with great potential for clinical application of multiresistant local infections [9] (Fig. 12.6).

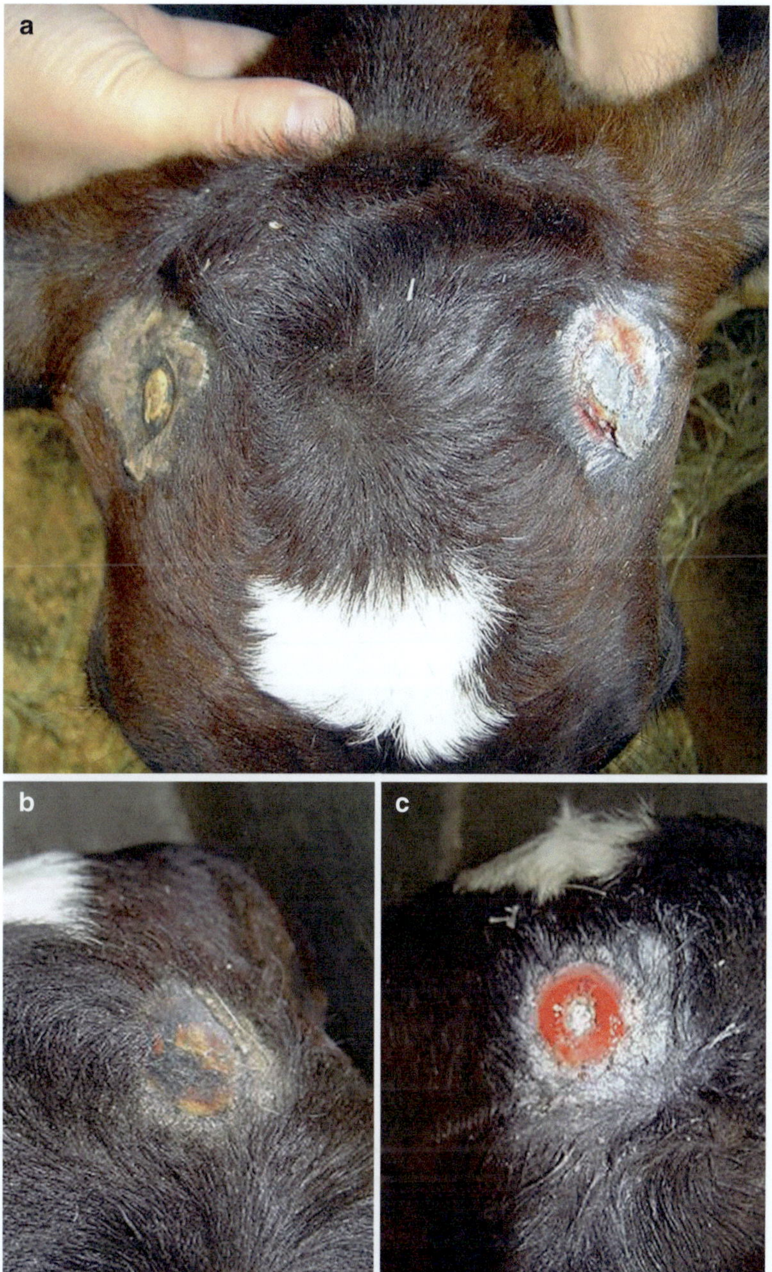

Fig. 12.5 Treatment of burn wounds after calf dehorning. (**a**) Lesion in the right side was daily treated with topic zinc oxide ointment (20 %) and in the left side was treated with PDT. MB was topically applied for 5 min followed by red laser irradiation. (**b**) Aspect of the left lesion treated by PDT after 15 days evidencing crust cover and absence of exudate or secondary infection. (**c**) Aspect of the right lesion treated on daily basis with zinc oxide ointment after 15 days. Note inflammation signs and absence of crust (Copyright image International Journal of Science Commerce and Humanities)

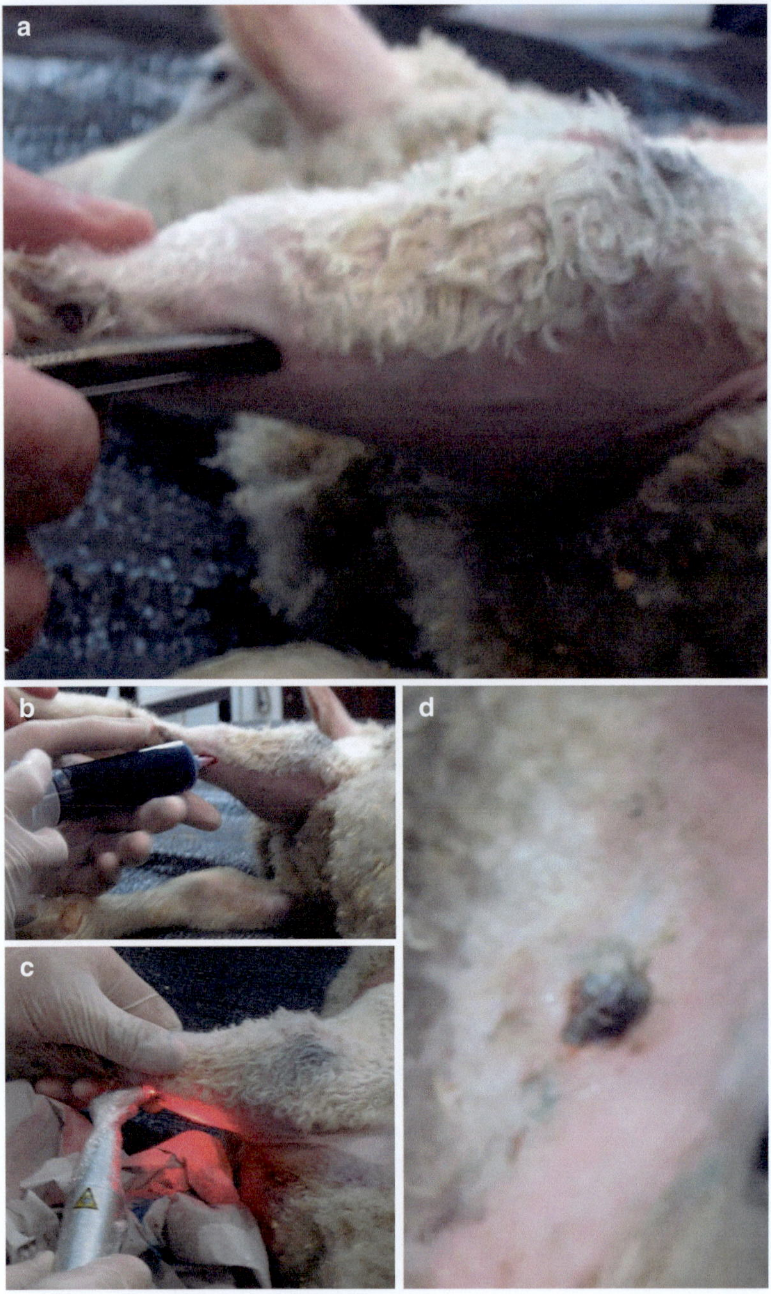

Fig. 12.6 Image of a cutaneous abscess in young bovine treated by PDT. (**a**) The abscess located on lateral side of the elbow joint was approximately 5 cm in diameter. (**b**) MB was applied topically for 5 min. (**c**) The lesion was irradiated using a diode red laser coupled with an optical fiber. (**d**) After 1 day, the lesion shows absence of inflammation and crust formation

In another case, a goat had myiasis and extensive tissue loss in the vulva and perianal region. After removing the larvae, the remaining wound was treated with PDT and aqueous solution MB (300 μM) followed by irradiation with diode laser (660 nm, 40 mW, and 0.4 J per point), distributed at points around the lesion. The complete reepithelialization took 15 days [10] (Fig. 12.7).

PDT to infectious diseases of the hoof and skin in horses and ruminants, respectively, also seems to be a good option (Figs. 12.8 and 12.9).

We treated two bovines with hoof lesions diagnosed as toe ulcer. The proposed treatment was performed applying MB aqueous solution (300 μM) followed by irradiation with red laser at $\lambda=660$ nm and energy of 8 J/point, twice a week until the complete resolution of the lesion (a total of eight sessions). The proposed therapy was effective for complete wound healing in an average period of 30 days [11] (Figs. 12.10 and 12.11).

Other studies reported the use of this technique for quick and effective treatment for ulcers of the sole and pinch, septic infiltration of the white line, distal interphalangeal septic arthritis, interdigital hyperplasia, necrotizing footpad lesions, and bovine digital dermatitis [12] (Figs. 12.12 and 12.13).

PDT also was successfully applied to treat caseous lymphadenitis abscesses in sheep. Ten sheep had their abscesses drained surgically and subsequently treated with MB (60 μM) followed by irradiation with red laser ($\lambda=660$ nm; energy/point, 4 J; power, 100 mW; irradiance/point, 3.3 W/cm^2; exposure time/point, 40 s; radiant exposure/point, 133.3 J/cm^2). All treated animals responded to PDT treatment with one or two treatments. After 1 week, no inflammatory signs and purulent exudate were observed. There was size reduction of all treated lymph nodes, returning to their normal sizes after 2 weeks. In this study, the lesions treated with PDT had a mean healing time of 15.3 days, time shorter than the treatments described in the literature [9] (Fig. 12.14).

12.4 Exotic, Zoo, and Wildlife Animals

The clinical practice for wild and exotic animals is an important field for veterinarians in search for precise information about health care of this species. The practitioner vet faces the challenge of a broad range of species with different needs passing through rehabilitation centers or in captivity in zoos and aquariums or even as an exotic pet.

The importance of conservation effort in counterpart to the human impact on the natural environment has been widely discussed over the last decades. Some species successfully adapted to the man, though others have had decreases in their populations. Thus, conservation programs challenge veterinarians who are facing species less known compared to domesticated ones.

These species have unique characteristics in terms of their behavior, environment where they thrive, feeding, and physiology, implying a great defy for professionals working in this area. New approaches for treatment and preventive care of diseases

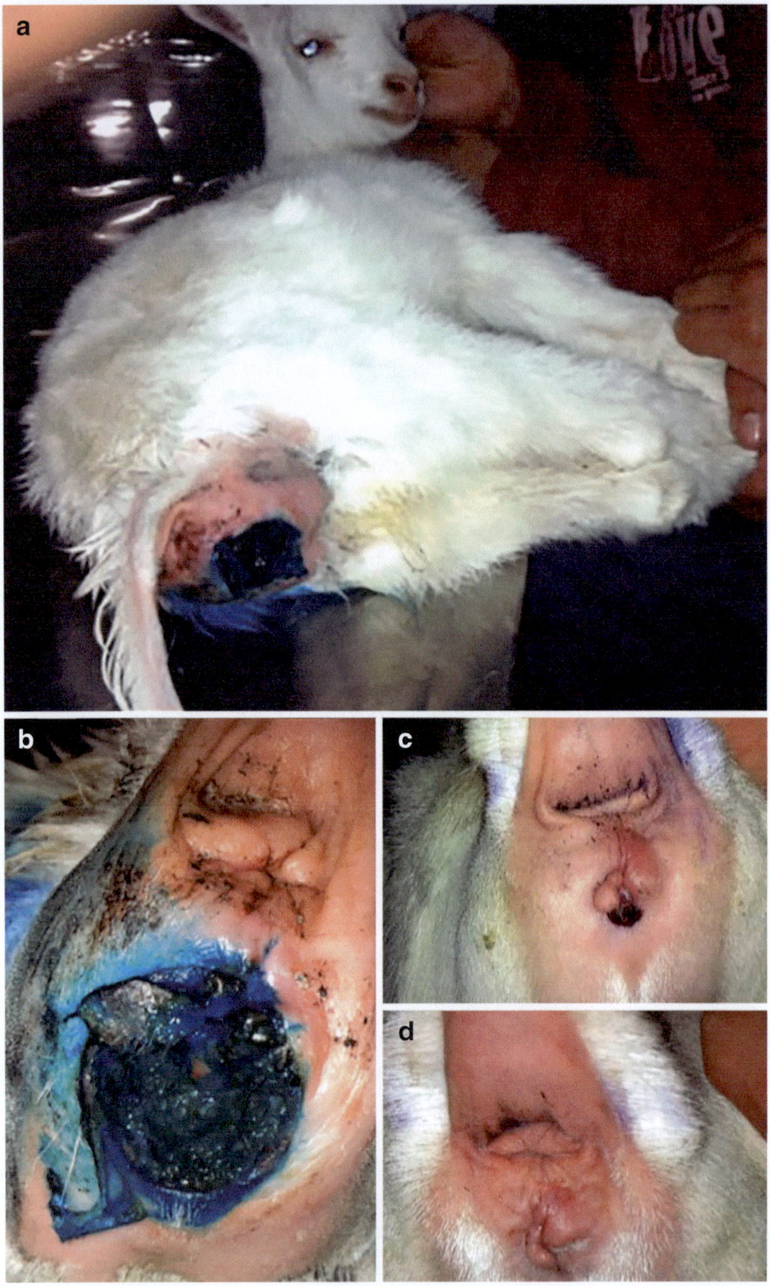

Fig. 12.7 Extensive lesion in the vulva of young goat treated by PDT. (**a**) Initial lesion evidencing tissue loss in the vulva and perianal region caused by myiasis. (**b**) A close of the lesion with abundant MB, which was topically applied in site for 5 min and irradiated with a red laser ($\lambda = 660$ nm). (**c**) Aspect of the lesion 7 days post PDT treatment, showing significant reduction of inflammation and good reepithelialization rate. (**d**) Aspect of the lesion after 15 days evidencing complete healing (Copyright image Acta Veterinaria Brasilica)

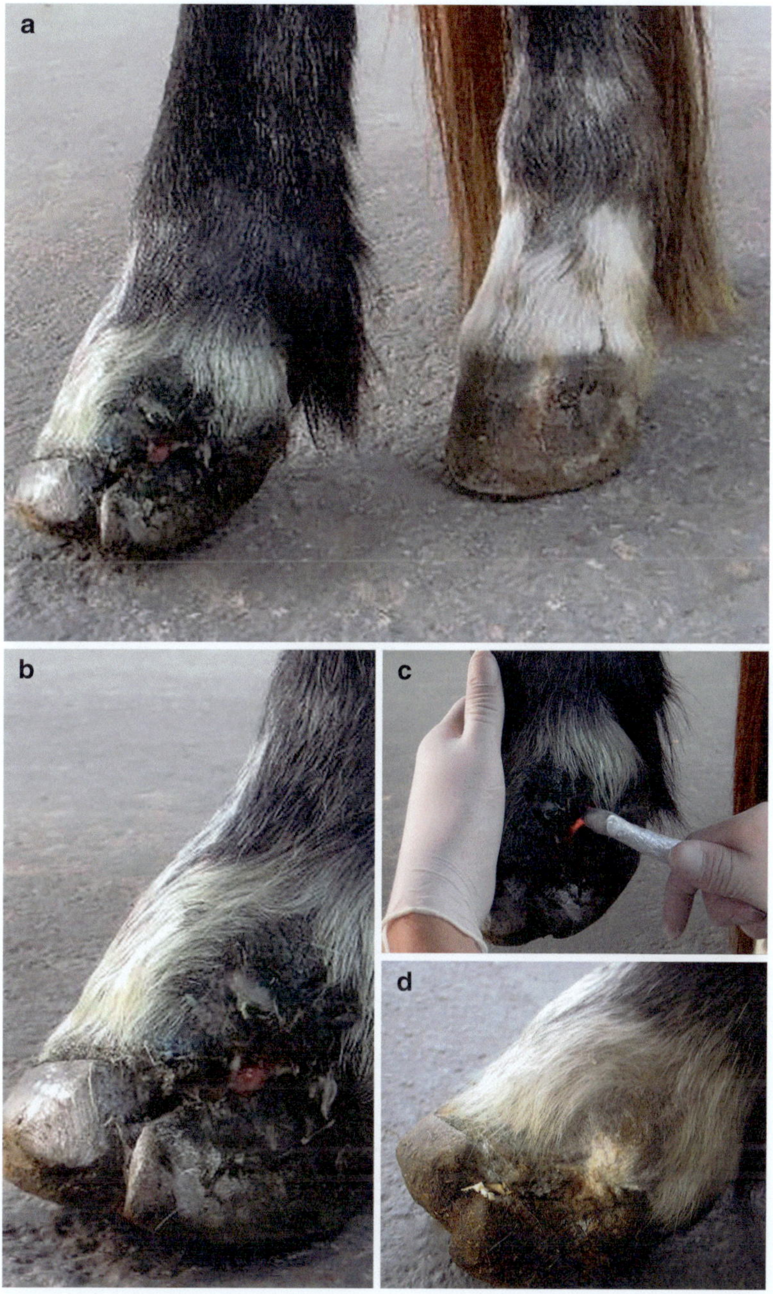

Fig. 12.8 (**a**, **b**) Infected hoof wound in horse; (**c**) PDT procedure performed in two consecutive days using 0.01 % MB topically applied for 5 min and irradiating with a red laser at $\lambda = 660$ nm, $P = 100$ mW, $E = 8.0$ J/point, and $\Delta t = 80$ s. (**d**) Complete wound healing after 1 week (Copyright image Journal of Continuing Education in Animal Science of CRMV-SP)

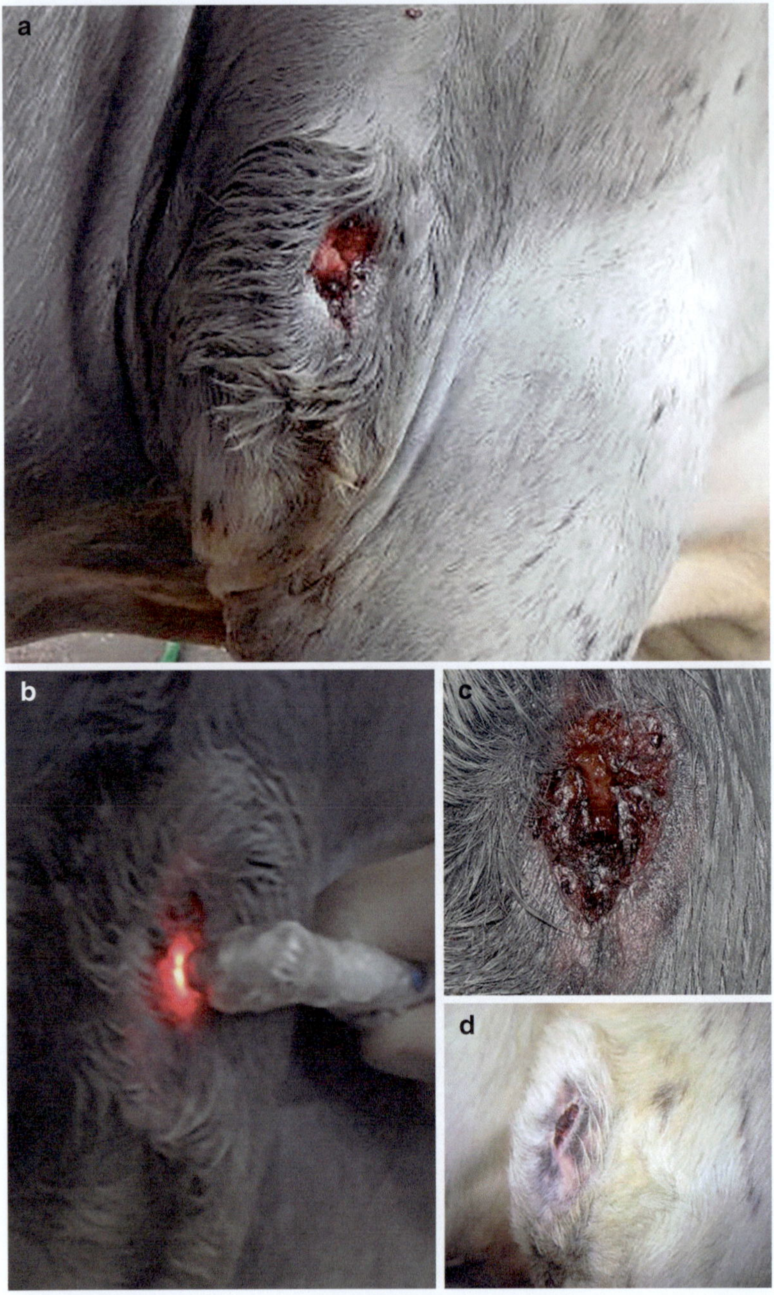

Fig. 12.9 (**a**) Abscess in shoulder region of a bovine. (**b**) Two PDT procedures in consecutive days using 0.01 % MB topically applied in site for 5 min and irradiated with a red laser ($\lambda=660$ nm) for 3 min ($P=100$ mW, $E=4$ J, fluence = 140 J/cm^2). (**c**) Aspect of the lesion 1 day after first PDT showing granulation tissue. (**d**) Aspect of the lesion after 1 week. The lesion is almost closed without inflammation signs

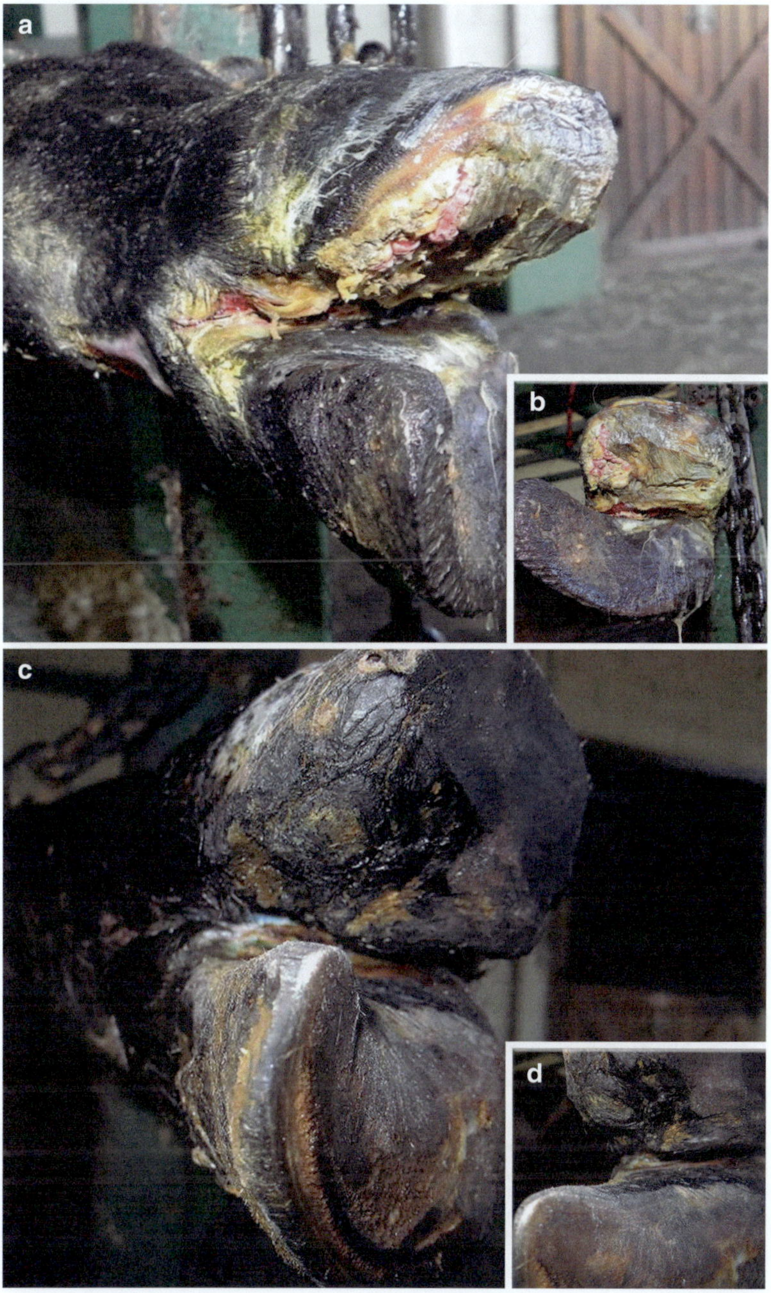

Fig. 12.10 Successful case of bovine toe ulcer treated by PDT. (**a**, **b**) Bovine toe ulcer in outer claw of left hind limb. (**c**, **d**) Complete wound healing 30 days post PDT (Copyright image European International Journal of Science and Technology)

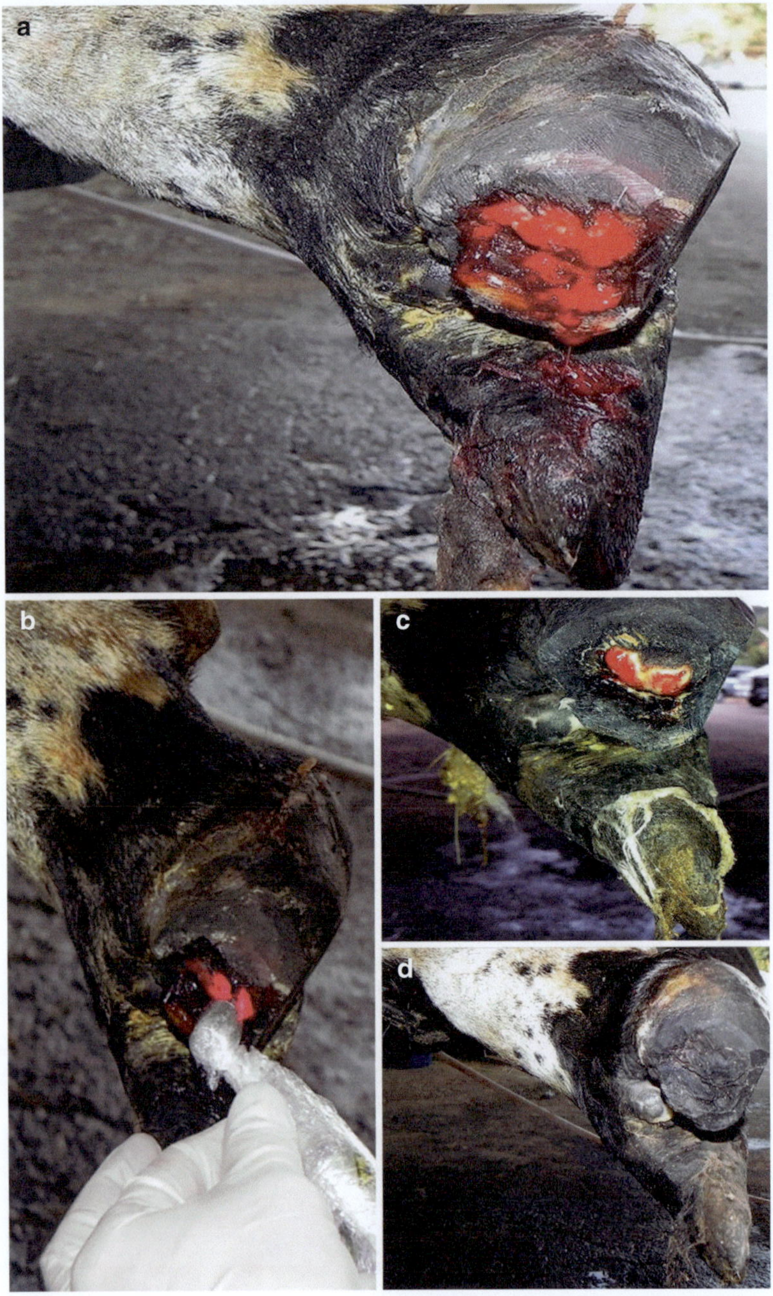

Fig. 12.11 Another case of bovine toe ulcer treated by PDT. (**a**) Bovine toe ulcer in inner claw of right hind limb presenting hemorrhage. (**b**) PDT was performed using MB topically applied in site for 5 min followed by red laser irradiation. (**c**) Partial healing after 21 days. (**d**) Complete wound healing after 30 days (eight PDT treatments) (Copyright image European International Journal of Science and Technology)

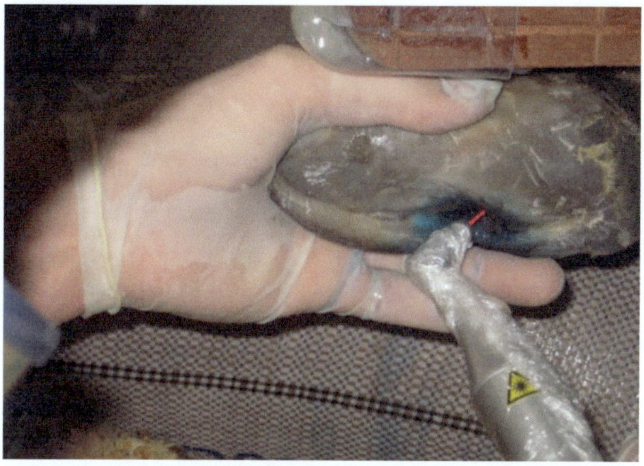

Fig. 12.12 Image showing an optical fiber coupled to the laser tip for bovine hoof abscess treatment. The optical fiber can deliver light into deep lesions as an abscess

are critical to the progress of this work front. Although the bulk of researches on PDT in veterinary medicine are to treat domestic animals, there is a great appeal for its clinical application in exotic, zoo, and wildlife medicine.

Some studies conducted by our team describe the application of PDT in the treatment of marine animals in rehabilitation and in captivity. In those studies, we use PDT for pododermatitis (bumblefoot) in Magellanic penguins (*Spheniscus magellanicus*). Bumblefoot is an ischemic injury similar to bedsore due to changes in the pattern of swimming of penguins in captivity, since they spend more time standing compromising the vascularization of the footpad, which allows MO to damage the dermis leading to erosion and ulcer formation [13].

At the first study, five Magellanic penguins (*S. magellanicus*) exhibiting Class III stage pododermatitis were treated with surgical excision on the necrotic fibrinous exudates [14]. At the remaining lesion, MB was applied (aqueous solution at a concentration of 300 μM), and after 5 min in the dark to allow MB uptake by the MO, the lesion was irradiated using a red laser emitting 100 mW at 660 nm, distributed in five equidistant points of 0.02 cm^2 that received 140 J/cm^2 each. The healing time varied according to the size of the wound. The average treatment time was approximately 8 weeks, and all five penguins presented a total lesion regression with no relapses during 6 months of follow-up (Fig. 12.15).

In another study, we compare outcomes in a group of captive *Spheniscus magellanicus* with bumblefoot lesions treated with PDT and antibiotics (ATB) [15]. Ten captive Magellanic penguins with preexisting stage III bumblefoot lesions were selected and randomly divided into PDT and ATB groups, each including 11 pelvic limb lesions. All animals underwent surgical debridement of lesions. In the ATB group, antibiotic ointment was applied topically three times a week, and systemic antibiotic and anti-inflammatory drugs were administered daily. In the PDT group, photodynamic therapy was applied three times a week without the use of topical or systemic medication using a red laser at $\lambda = 660$ nm power of 100 mW during 40 s applied at 1 cm equidistant points as many as necessary to cover the injury and delivering an energy of 4 J/point. Lesion areas were photographed, and swabs were

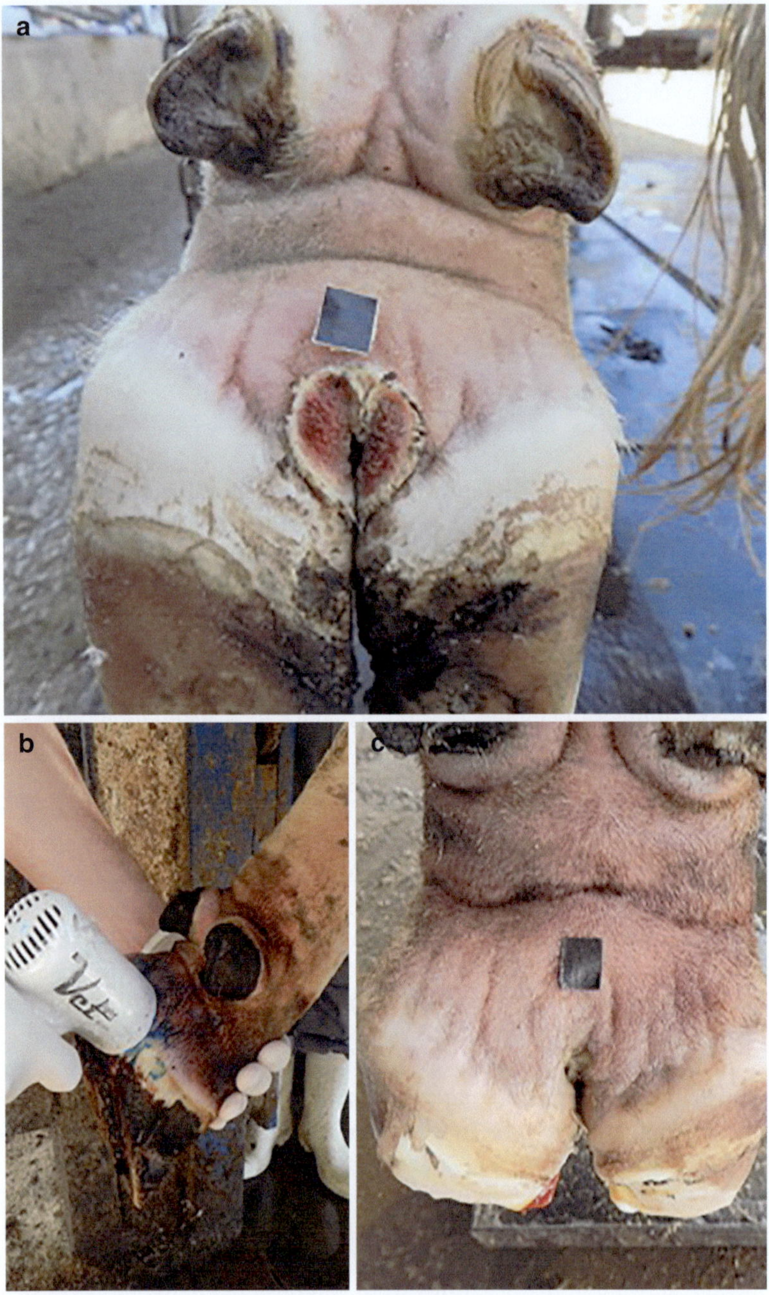

Fig. 12.13 Bovine digital dermatitis treated by PDT. (**a**) Classical ulcerative bovine digital dermatitis lesion. (**b**) PDT was performed every 15 days with 0.01 % MB in site for 5 min and irradiation using a cluster of red LEDs ($\lambda = 655$ nm, $P = 105$ mW, $E = 12$ J). (**c**) Complete wound healing after 30 days (two treatments)

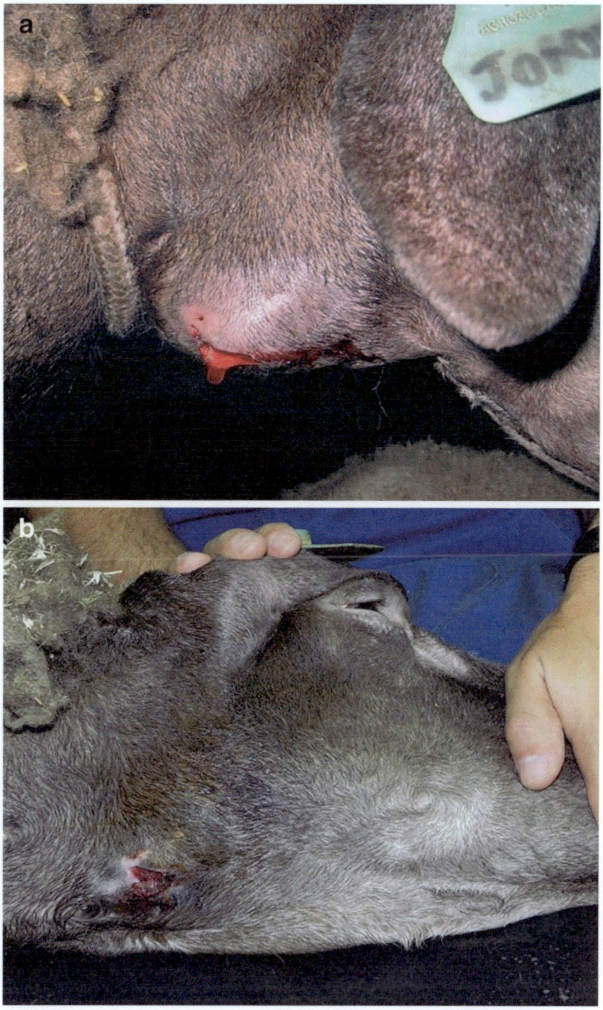

Fig. 12.14 Caseous lymphadenitis abscess in mandibular lymph node treated by PDT. (**a**) Aspect of the lesion after surgical drainage. (**b**) Wound healing after 3 days from PDT session. Note the reduction of inflammation and lesion area

collected for culture and sensitivity, on the first day and every 14 days for a total of 84 days. The culture test showed 11 bacterial species that varied along the time. The four species of bacteria that showed higher resistance to the antibiotics were selected to determine resistance to PDT in vitro. Comparing the healing rates between the groups, there was a significant difference for PDT with 63.64 % of healed lesions compared with 9.09 % for ATB group. Regarding the duration of treatment until complete healing, there was a significant difference between groups. The PDT group showed a variation between 28 and 70 days, the average being 42 days. In the ATB group, the only lesion healed took 70 days to complete healing.

The main challenge in the treatment of bumblefoot is the constant contamination and compression of the injury accentuating the ischemic process. Penguins do not perch, so they are constantly in contact with feces and urine, and it is very difficult

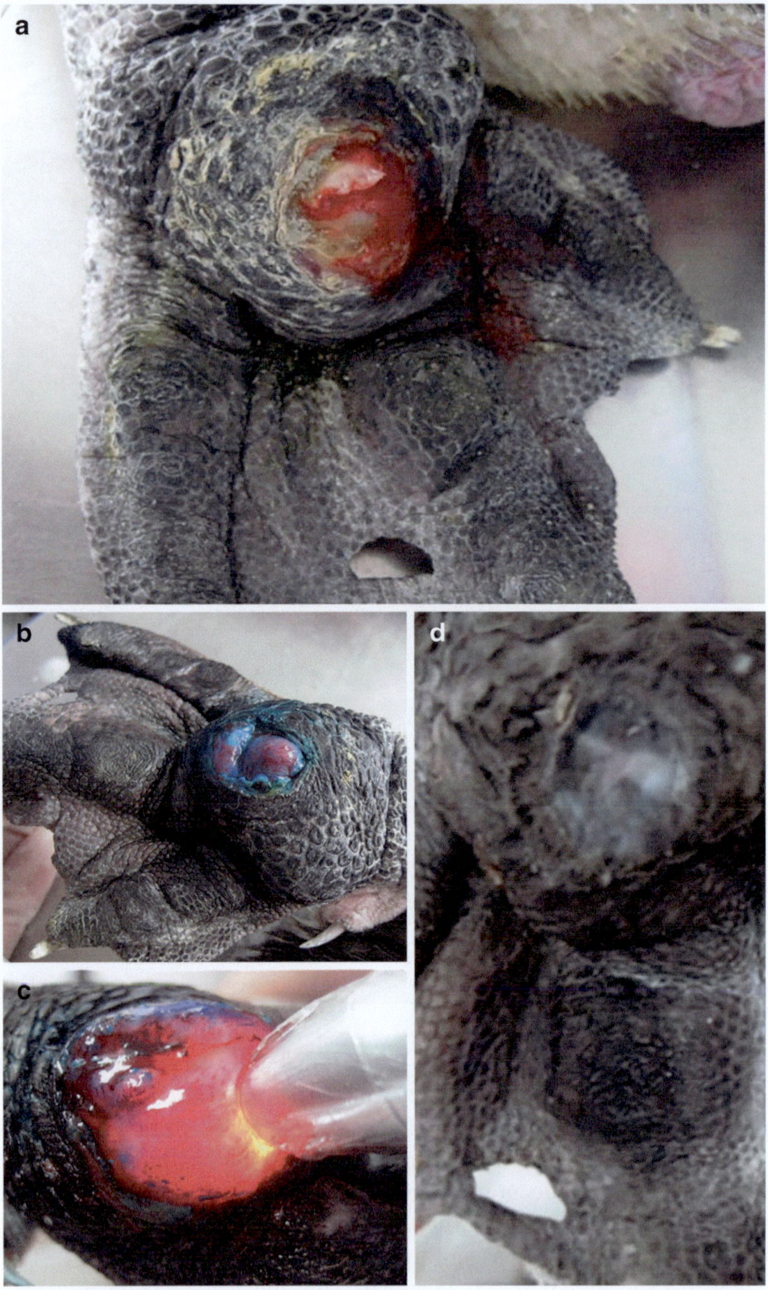

Fig. 12.15 (**a**) Stage III footpad dermatitis (bumblefoot) in left hind limb in Magellanic penguin (*Spheniscus magellanicus*). (**b**) MB applied after surgical debridement and left in site for 5 min. (**c**) Irradiation using a red laser in five equidistant points. (**d**) Complete wound healing after 15 days (two PDT sessions)

to avoid recontamination of wounds even if environmental hygiene is constantly maintained. PDT can be applied directly to the infected area, selectively destroying a large number of MO without causing tissue destruction. One of the advantages of PDT over antibiotics, which depend on bioavailability and effectiveness against bacterial contaminants at the time, is its broad-spectrum and localized action of PDT. Thus, PDT is appropriate for bumblefoot treatment because it does not develop resistance, and topical application obviates concerns about bioavailability.

Other avian species are prone to develop bumblefoot, such as birds of prey and waterfowls [16–19]. The PDT approach is a promising field of study for the treatment of these bedsore-like lesions.

Our team also tested the PDT to disinfect fracture sites in the carapace of sea turtles in rehabilitation. The dermis of these animals is ossified and the epidermis is modified in corneal tissue [20]. The shell is a natural barrier of the organism, which isolates the internal components of the external environment [21]. When this barrier is partially or totally destroyed, the animal survival may be compromised.

One green turtle was being treated for 6 months with antibiotic ointments on a fracture located on the third, fourth, and fifth vertebral shields and the other two in the fourth right costal shield and eleventh marginal right shield. The bones were not fractured. The fractured areas began to be treated exclusively with PDT (MB aqueous solution concentration of 60 μM, red laser at $\lambda=660$ nm with 100 mW of power with an energy density of 180 J/cm^2 per point). Treatments were performed with an interval of one application per week until complete regression of the lesion. The wound healing and keratinization were obtained after four treatments (28 days) (Fig. 12.16).

Overall, we believe that wild animals with infected wounds are suitable for treatment by PDT. We have observed in clinical practice good results also in species such as vulture (*Coragyps atratus*) (Fig. 12.17), bush dog (*Speothos venaticus*) (Fig. 12.18), golden-headed lion tamarin (*Leontopithecus chrysomelas*), and other primates of the genus *Callithrix* (Fig. 12.19).

The wild animal medicine seems to open a huge range for developing research involving PDT, because little is known about the different species, metabolism, diseases, and their treatments. About 10 years ago, Lucroy raised the possibility of applying PDT on the treatment of ulcerative infectious stomatitis in reptiles [22].

Stomatitis mainly occurs in immunosuppressed animals due to stress during the captive adaptation period. Immunosuppression causes the exchange of oral microbiota of bacteria gram-positive to gram-negative. Common symptoms of stomatitis are inappetence, regurgitation, petechiae, excessive production of saliva, and accumulation of caseous materials along the dental arcade. If not treated, stomatitis could lead to pneumonia, osteomyelitis, or septicemia. Generally, the administration of broad-spectrum antibiotic is the best choice, often for a long period; however, its use does not guarantee treatment success [23, 24].

We investigate PDT to treat three captive boid snakes with infectious stomatitis. Mix infections were diagnosed by microbiological test. To perform PDT, the snakes were kept open mouth with an anatomical tweezer, and caseous material was removed followed by application of 1 mL of MB dye (0.01 %) direct to the lesions. After 5 min of PIT, we irradiated the lesions with a red laser emitting at $\lambda=660$ nm, 100 mW of

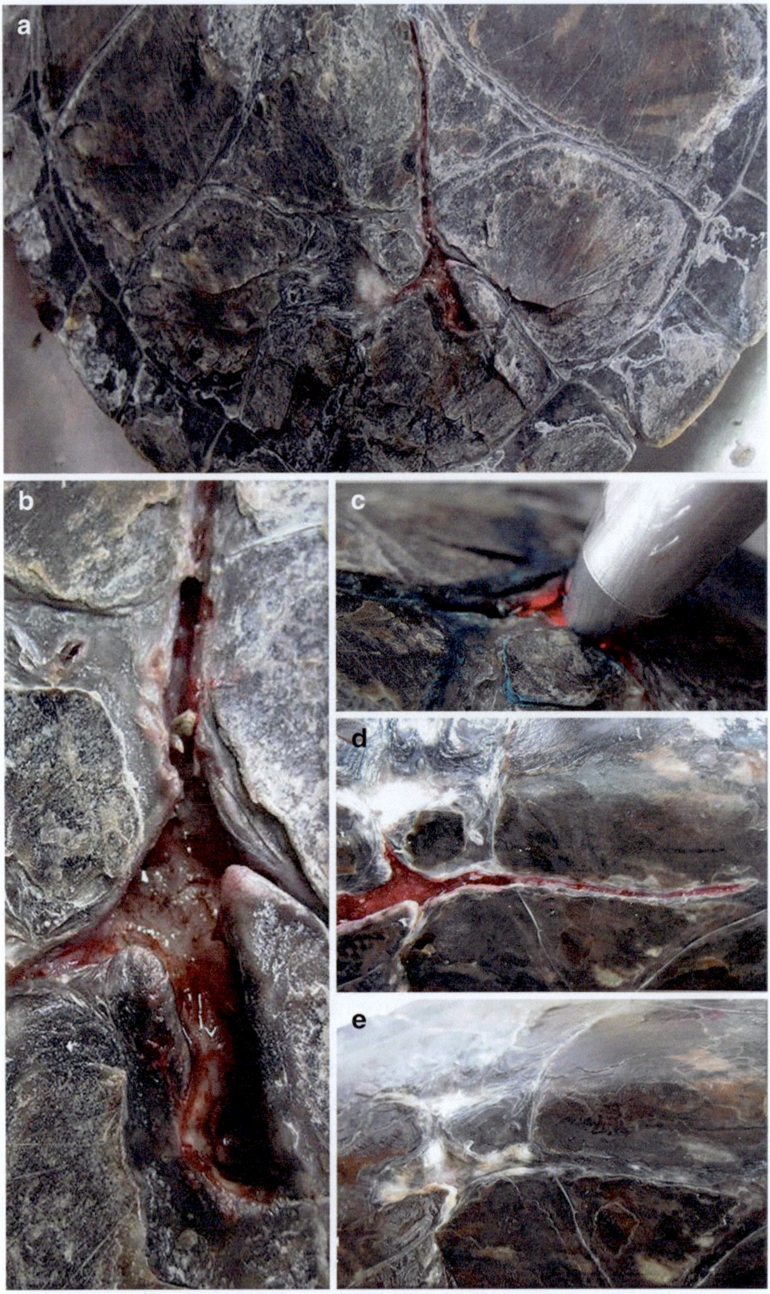

Fig. 12.16 Image of infected fracture in the carapace of green turtle (*Chelonia mydas*) treated by PDT. (**a**, **b**) Initial fracture aspect located on third, fourth, and fifth vertebral shields, fourth costal shield, and 11th marginal shield showing exudate and tissue necrosis. (**c**) MB was applied using a syringe and left in site for 5 min followed by red laser illumination. (**d**) Aspect of the lesion after 7 days (one PDT session) evidencing absence of infection and reduction of fractured area. (**e**) Aspect of the lesion after 15 days (two PDT sessions); the fracture is almost consolidated without inflammation signs

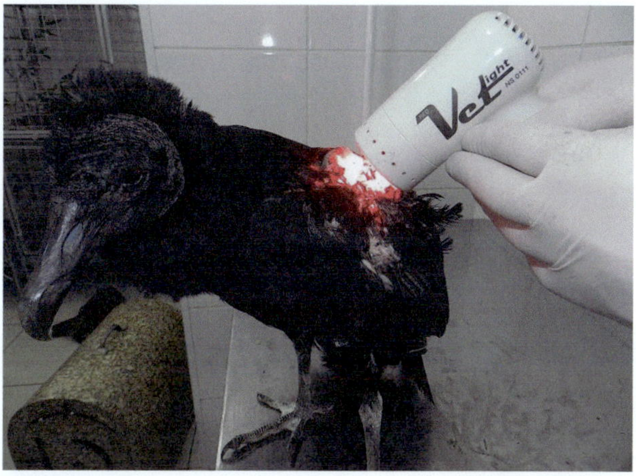

Fig. 12.17 PDT procedure on infected wound in vulture wing (*Coragyps atratus*) using a cluster of red LEDs. LED devices are useful for extensive lesions

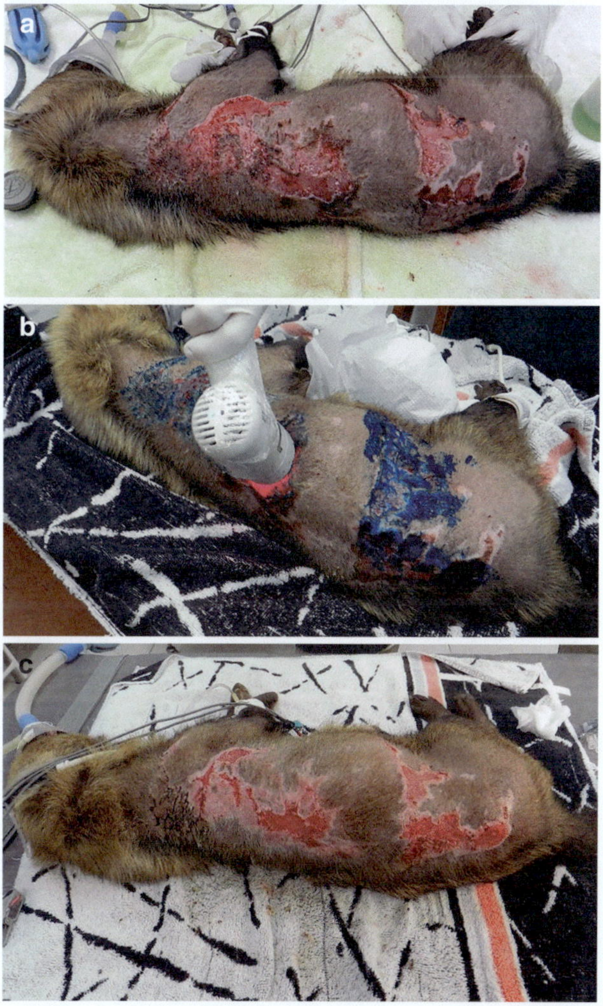

Fig. 12.18 An infected burn wound in a bush dog (*Speothos venaticus*) treated by PDT. (**a**) Initial aspect of lesion presenting extensive tissue loss, inflammation, and necrosis. (**b**) PDT with 0.01 % MB in gel formulation in site for 5 min and irradiation using a cluster of red LEDs ($\lambda = 655$ nm, $P = 105$ mW, 8 J/point). (**c**) After 7 days (one PDT session), the lesion shows absence of necrosis, good reepithelialization, and granulation tissue formation

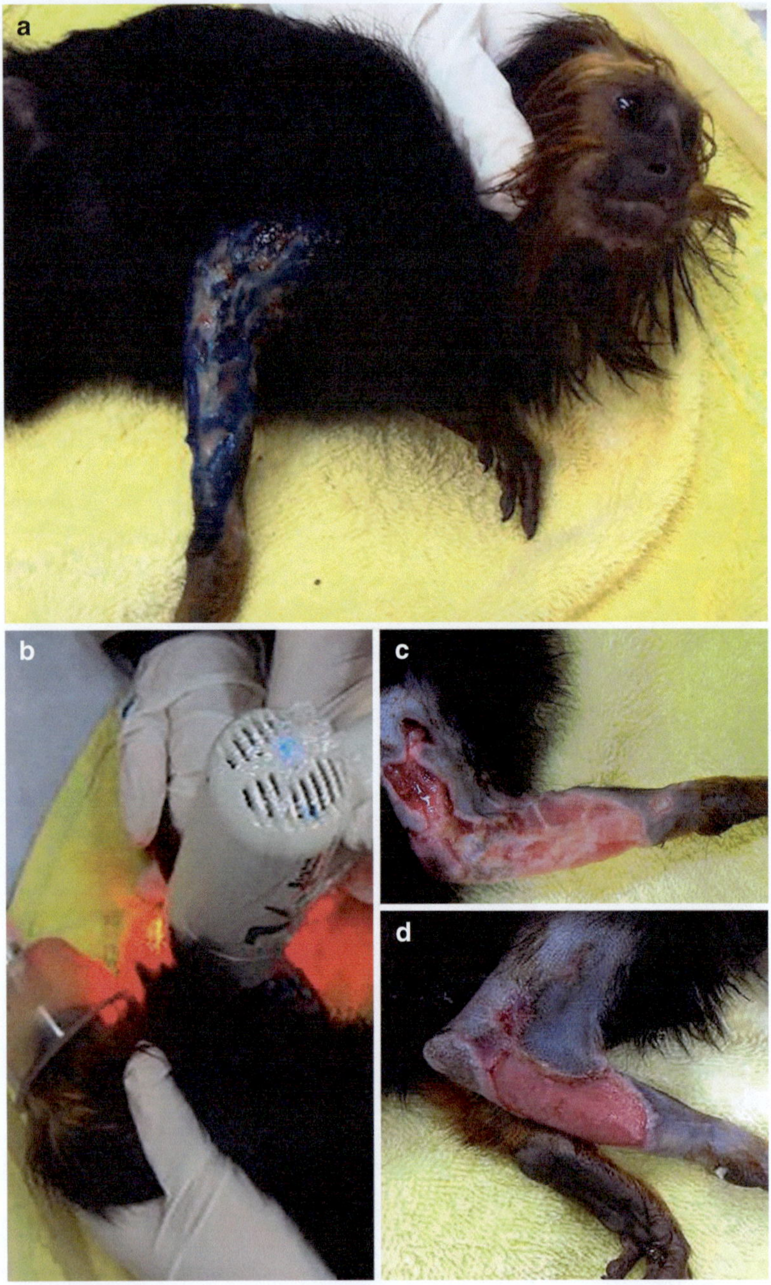

Fig. 12.19 Image of a self-inflicted wound in a golden-headed lion tamarin (*Leontopithecus chrysomelas*) treated by PDT. (**a**) Aspect of the lesion covered with 0.01 % MB gel-formulated. (**b**) Irradiation procedure using a cluster of red LEDs ($P = 105$ mW, $\lambda = 655$ nm, 8 J/point). (**c**, **d**) Aspect of the lesion after 3 and 7 days, respectively. Note the absence of infection and good reepithelialization rates in a short time

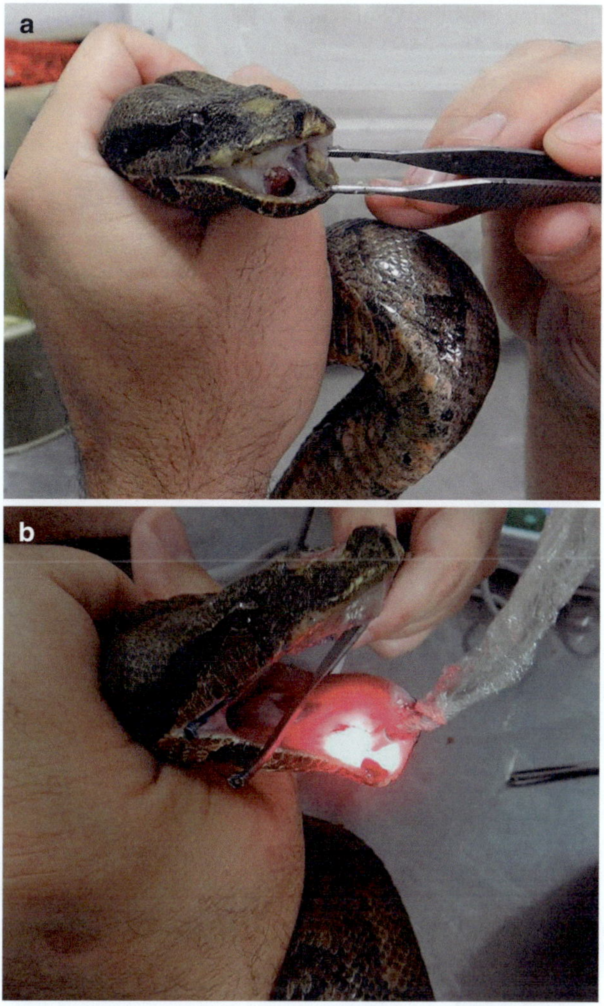

Fig. 12.20 PDT to treat infectious stomatitis in a boid snake (*Boa constrictor*). (**a**) Classical stomatitis lesion. (**b**) Irradiation procedure after MB topically applied. It is possible to keep snakes' mouth open with anatomical tweezers to perform laser irradiation

output power and irradiance of 3.5 W/cm^2, 80 s per point delivering a radiant exposure of 280 J/cm^2, and energy of 8 J. PDT proved to be an interesting alternative for infectious stomatitis in snakes avoiding antibiotics for long periods (Fig. 12.20).

Sea turtle fibropapillomatosis is a disease marked by proliferation of benign but debilitating cutaneous fibropapillomas and occasional visceral fibromas. Lackovich et al. concluded that a chelonian herpesvirus is regularly associated with fibropapillomatosis. It is not merely an incidental finding in affected turtles; it also was concluded that herpesvirus was detectable only within or close to tumors [25].

Our team got success using PDT in five green turtles (*Chelonia mydas*) with multiple fibropapillomatosis tumors [26]. The lesions were treated by two PDT applications within a 15-day interval. PDT consisted of two injections of 0.5 mL intralesional 300 μM MB in the base of all tumors, followed by 5 min of PIT and then a red laser operating at 100 mW of output power, λ=660 nm, and 160 s/point (energy of 16 J and energy density of 560 J/cm^2 per point) was used. After 7 days, all lesions were dark blue in color and swollen with firm consistency, followed by a soft consistency in the next 7 days with classical macroscopic characteristics of tissue necrosis. During the succeeding 14 days, lesions showed partial loss of adherence to the skin, became partially or totally detached from the skin, and were easily removed using tweezers (Fig. 12.21).

This chapter addressed successful cases of antimicrobial PDT in veterinary medicine. We hope to encourage veterinarians to new clinical studies to promote PDT as a feasible approach to manage infectious diseases with a good compliance by the animals.

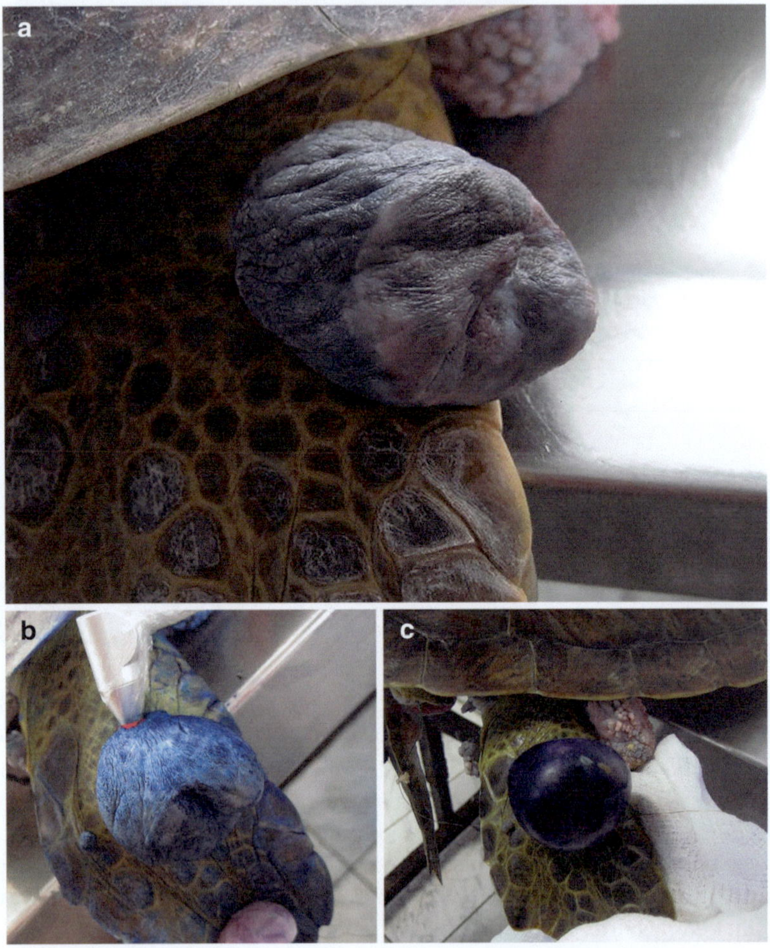

Fig. 12.21 Fibropapillomatosis in green turtle (*Chelonia mydas*) treated by PDT. (**a**) Aspect of one tumor before treatment. (**b**) Tumor was treated by two PDT using 0.5 mL MB intralesional around the base of the tumor and irradiation with a red laser. (**c**) Aspect of the tumor after 1 week. Observe that tumor is swollen due to intense inflammatory process. (**d**, **e**) Tumor after 2 and 3 weeks from first PDT. Note that the tumor shows macroscopic aspects of necrosis

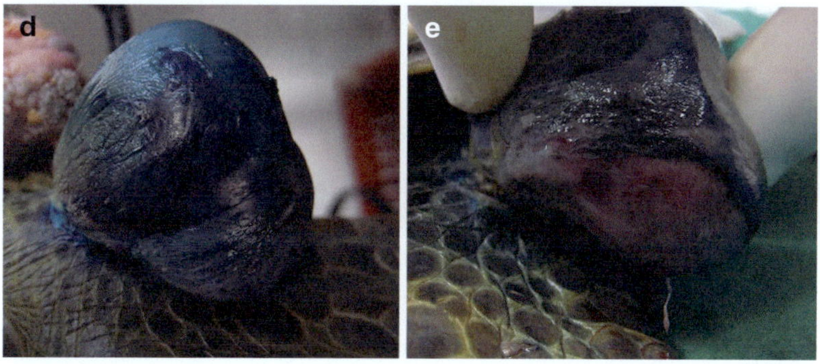

Fig. 12.21 (continued)

References

1. Fabris C, Soncin M, Corsi F, Cattin I, Jori G. Treatment of spontaneous piodermitis in dogs by PDT using porphyrin-type photosensitisers. Photodiagnosis Photodyn Ther. 2011;8(2):138.
2. Hayek R, Ricardo R, Araújo NS, Ney S, Gioso MA, Ferreira J, et al. Comparative study between the effects of photodynamic therapy and conventional therapy on microbial reduction in ligature-induced peri-implantitis in dogs. J Periodontol. 2005;76(8):1275–81.
3. Sellera FP, Nascimento CL, Azedo MR, Sellera DP, Cecília A, Aranha C, et al. Photodynamic therapy in the treatment of canine oral papillomatosis. Int J Sci Commer Hum. 2013; 1(8):23–7.
4. Pimprikar GD, Noe BL, Norment BR, Heitz JR. Ovicidal, larvicidal, and biotic effects of xanthene derivatives in the house fly, Musca domestica L. Environ Entomol. 1980;9(6):785–8.
5. Hawkins JA, Healey MC, Johnson-Delivorias MH, Heitz JR. The effect of erythrosin B on infective larvae of bovine gastrointestinal nematodes. Vet Parasitol. 1984;16(1–2):35–41.
6. Hawkins JA, Johnson-Delivorias MH, Heitz JR. Photodynamic action of erythrosin B as a toxic mechanism for infective larvae of bovine gastrointestinal nematodes. Vet Parasitol. 1986;21(4):265–70.
7. Healey MC, Smith MB, Smith LD. The phototoxic effect of erythrosin B on third-stage larvae of gastrointestinal nematodes in sheep. Vet Parasitol. 1992;43(3–4):249–57.
8. Sellera FP, Azedo MR, Silva LCB a, Seino CH, Gargano RG, Batista CF, et al. Photodynamic therapy as a promising treatment of burn wounds after calf dehorning. Int J Sci Commer Humanit. 2013;1(8):28–32.
9. Sellera FP, Gargano RG, Libera AMMPD, Benesi FJ, Azedo MR, de Sá LRM, et al. Antimicrobial photodynamic therapy for caseous lymphadenitis abscesses in sheep: report of ten cases. Photodiagnosis Photodyn Ther. 2016;13:120–2.
10. Sellera FP, Barbosa SB, Gargano RG, Benesi FJ, Pogliani FC. Photodynamic therapy treatment of wound caused by myiasis in the goat vulva – case report. Acta Vet Bras. 2014;8(1):74–7.
11. Sellera FP, Gargano RG, Azedo MR, Benesi FJ, Lopes LA, Pogliani FC. Antimicrobial photodynamic therapy as an adjuvant treatment of toe ulcer in cattle. Eur Int J Sci Technol. 2013;2(10):98–104.
12. Sellera FP, Souza ASL, Ferreira VG, Alves MBR, Azedo MR, Benesi FJ, et al. Photodynamic therapy in cattle hoof diseases. In: XXVII World Buiatrics: Lisboa; 2012. p. 267–8.
13. Erlacher-Reid C, Dunn JL, Camp T, Macha L, Mazzaro L, Tuttle AD. Evaluation of potential variables contributing to the development and duration of plantar lesions in a population of aquarium-maintained African penguins (Spheniscus demersus). Zool Biol. 2011;31(3): 291–305.

14. Sellera FP, Sabino CP, Ribeiro MS, Fernandes LT, Pogliani FC, Teixeira CR, et al. Photodynamic therapy for pododermatitis in penguins. Zool Biol. 2014;33(4):353–6.
15. Nascimento CL, Simões MR, Sellera FP, Dutra GHP, Simões A, Teixeira CR. Comparative study between photodynamic and antibiotic therapies for treatment of footpad dermatitis (bumblefoot) in Magellanic penguins (Spheniscus magellanicus). Photodiagnosis Photodyn Ther. 2015;12(1):36–44.
16. Cooper JE, Max RA, Mbassa GK. Health studies on a group of captive helmeted guineafowl (Numida meleagris) in Tanzania. Avian Pathol. 1996;25(1):135–45.
17. Cooper JE. Surgery of the foot in falcons: a historic operation. Ann R Coll Surg Engl. 1980;62(6):445–8.
18. Cooper JE. Birds of prey. In: Cooper JE, editor. Birds of prey: health & disease. 3rd ed. Oxford: Blackwell Science Ltd; 2002. p. 121–31.
19. Cooper JE, Needham JR. The starling (Sturnus vulgaris) as an experimental model for staphylococcal infection of the avian foot. Avian Pathol. 1981;10(3):273–9.
20. Divers S. The structure and diseases of the chelonian shell. Certain aspects of the veterinary care of chelonian. Sevenoaks. 1996;4(3):10–8.
21. Kaplan M. Turtle and tortoise shell. Los Angeles. Herpetol Care Collect. 2002:78–84.
22. Lucroy M. Photodynamic therapy: potential applications for exotic animal oncology. Semin Avian Exot Pet Med. 2005;14(3):205–11.
23. Kaplan M, Jereb R. Ulcerative stomatitis (mouthrot) in reptiles. J Wildl Rehabil. 1995;18(2):13.
24. Mehler S, Bennett R. Upper alimentary tract disease. In: Mader D, editor. Reptile medicine and surgery. 2nd ed. St. Louis: Elsevier/Saunders; 2006. p. 924–30.
25. Lackovich JK, Brown DR, Homer BL, Garber RL, Mader DR, Moretti RH, et al. Association of herpesvirus with fibropapillomatosis of the green turtle Chelonia mydas and the loggerhead turtle Caretta caretta in Florida. Dis Aquat Organ. 1999;37:89–97.
26. Sellera FP, Sabino CP, Fernandes LT, Pogliani FC, Teixeira CR, Dutra GHP, et al. Green turtle (Chelonia mydas) cutaneous fibropapillomatosis treatment by photodynamic therapy. Mar Turt Newsl. 2014;142:1–31.

Chapter 13
Other Practices in PDT

Fábio Parra Sellera, Fabio Celidonio Pogliani, and Caetano Padial Sabino

Abstract In addition to clinical PDT applications regarding antimicrobial and antineoplastic activity, photodynamic reactions have also been used in several other practices such as for fish tank decontamination, water treatment, antiangiogenic therapy for age-related macular degeneration, decontamination of surfaces, and even inactivation of pathogens for blood transfusion. Nowadays, not all potentials of photodynamic reactions are commercially available yet, but they definitely deserve to be highlighted in this chapter as alternative applications of photodynamic reactions in veterinary medicine.

13.1 Introduction

In the previous chapters, we addressed the major uses applications of photodynamic therapy (PDT), i.e., for cancer and infections. However, PDT has been used in many other practices such as to kill fish parasites, water disinfection, age-related macular degeneration, surface decontamination, atherosclerosis, and pathogen eradication of blood for transfusion. Below we discuss some alternatives to use PDT in Veterinary sciences.

F.P. Sellera (✉) • F.C. Pogliani
Department of Internal Medicine, School of Veterinary Medicine and Animal Science, University of São Paulo, Butantã, São Paulo, SP, Brazil
e-mail: fsellera@usp.br; fabiocp@usp.br

C.P. Sabino
Department of Microbiology, Institute for Biomedical Sciences, University of São Paulo, São Paulo, SP, Brazil

Department of Clinical Analysis, School of Pharmaceutical Sciences, University of São Paulo, São Paulo, SP, Brazil

Center for Lasers and Applications, Nuclear and Energy Research Institute, National Commission for Nuclear Energy, São Paulo, SP, Brazil

Princess Margaret Cancer Institute/University of Toronto, Department of Medical Biophysics, Toronto, ON, Canada
e-mail: caetanosabino@gmail.com

F.P. Sellera et al. (eds.), *Photodynamic Therapy in Veterinary Medicine: From Basics to Clinical Practice*, DOI 10.1007/978-3-319-45007-0_13

tdown for Ophthalmology

a highly complex region that concentrates large amounts of ponsible for our high-definition central vision. Choroid is the blood vessels permeate to supply nutrients and oxygen for all choroidal neovascularization (CNV) is the leading cause of blind- humans and can occur as a consequence of pathologic myopia, chronic tion, and trauma but is most commonly associated with age-related macu- generation (AMD). It is currently believed that a number of factors may be sociated with the etiology of AMD. Fair-skinned individuals with blue or green eyes seem to be more susceptible. In addition, other factors such as excessive exposure to sunlight, smoking, and high-fat diet can also contribute to its higher incidence. As a common consequence, lipid deposits (drusen) build up in the cells of the retinal-pigmented epithelium causing chronic inflammation that eventually leads to atrophy and scarring of the retina. About 90 % of affected patients have a less severe form referred to as dry or non-exudative. The exudative, or wet, AMD is primarily responsible for the rapid devastation of central visual associated with macular degeneration [1]. This is the most severe form characterized by the development of abnormal blood vessels in the macula. These vessels are often leaky and bleed causing macular edema due to consequent inflammation. The macular edema causes local hypoxia and induces proliferation of further abnormal blood vessels stimulated by vascular endothelial growth factor (VEGF). Intravitreous administration of anti-VEGF monoclonal antibodies can be effective against the less aggressive CNV disorders. However, immunotherapy is less effective against more severe disorders, such as those involving formation of polyps or malignant lesions [1].

In recent decades, PDT has become a robust ally to selectively destroy functional blood vessels in diseases involving neovascularization (e.g., tumor and retinal angiogenesis) [2]. In fact, the most financially successful application of PDT was dedicated to the treatment of CNV disorders. The standard FDA-approved protocol employs benzoporphyrin derivatives (verteporfin) encaged in liposomes as photosensitizer [3]. After intravenous injection, the PS-loaded liposomes bind to low-density lipoproteins (LDL) as it is transported through the blood stream. The pathologic endothelial cells of CNV express up to tenfold increase in LDL-receptors and, therefore, uptake much greater amounts of LDL loaded with photosensitizers than healthy endothelial cells. When photosensitized endothelial cells are illuminated using a low-power laser, reactive oxygen species (ROS) are generated causing oxidative damage to cytoskeleton. Oxidized amino acids alter protein folding and can induce dramatic changes in three-dimensional structure of a protein complex. Hence, cytoskeletal changes caused by oxidation make the affected endothelial cells contract disrupting intercellular junctions. This process exposes the vascular basal membrane triggering the release of various clotting factors which lead to platelet activation. Platelets then attach to the exposed collagen of basal membrane, induce formation of fibrin clots and vasoconstriction, and finally obstruct blood flow through the targeted vessel [3]. Figure 13.1 illustrates PDT for age-related macular degeneration.

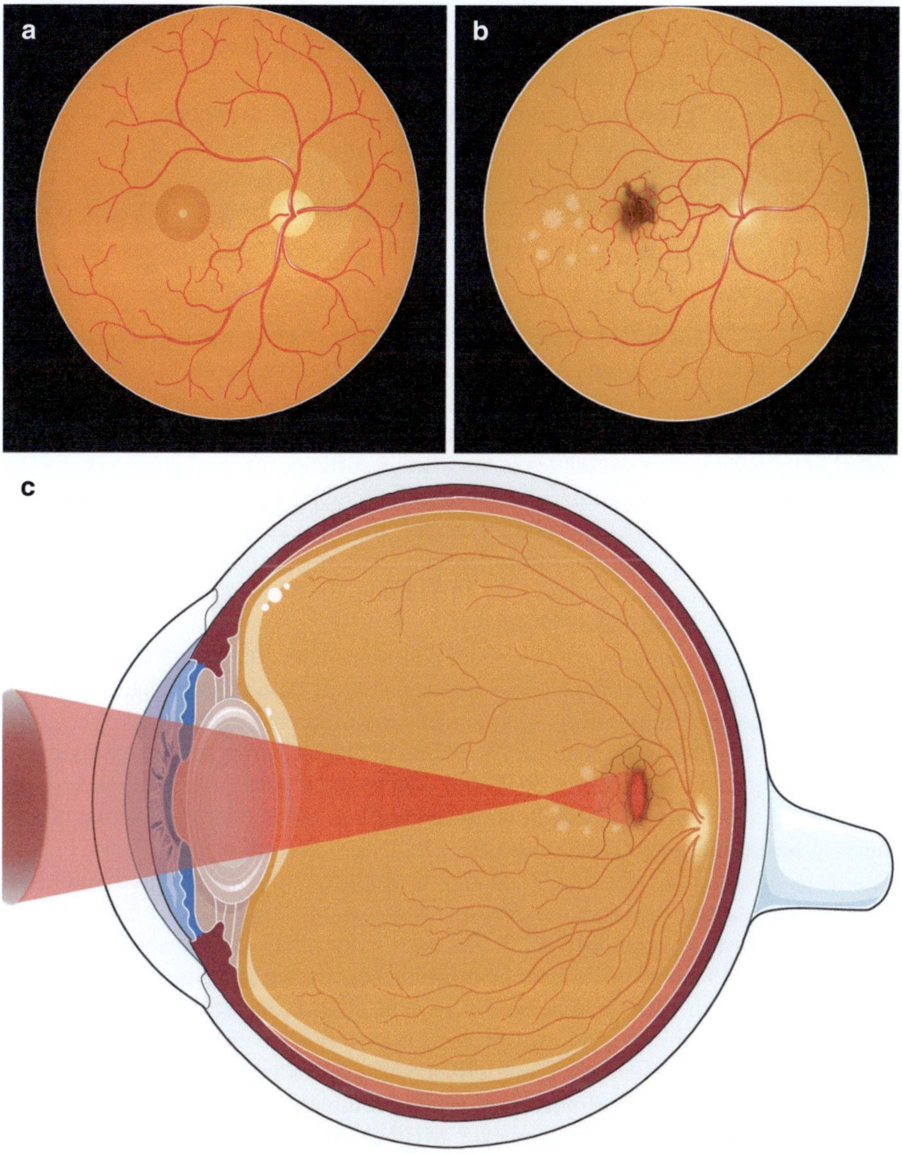

Fig. 13.1 Schematic illustration of a healthy retina (**a**) and a retina affected by age-related macular degeneration (**b**). How it is irradiated to perform PDT can be noticed in (**c**) [4]

The standard protocol used for most cases of CNV, including for AMD, recommends illumination by 689 nm laser onto macular surface at 600 mW/cm^2 for 87 s (i.e., total light dose of 50 J/cm^2). Treatment can be repeated every 3 months for periods beyond 2 years, if CNV leakage is still observed. Nowadays, canine models are being used to the understanding of retinal disease mechanisms for the development of new therapies in veterinary medicine [5, 6]. Beyond similarities in ocular anatomy and the close genotype-phenotype correlation [5], the literature still lacks on studies regarding PDT and macular dystrophy in dogs. Likewise, veterinarians could use such positive outcomes obtained in humans to propose studies using PDT for retinal degenerative disorders in similarly affected animals.

13.3 Blood Disinfection

The aim of PDT for blood disinfection is to avoid disease transmission through blood transfusion [7]. Although the number of blood transfusions in animals is far less than humans nowadays, the importance of this route in dissemination of several diseases should always be highlighted. The sensibility of available techniques to detect pathogens in blood samples is not always sufficient to guarantee transfusion safety. Bacteria, viruses, protozoans, or yeasts can be present in small amounts that are sufficient to avoid detection but yet capable to infect an acceptor. Therefore, the development of broad-spectrum antimicrobial technologies capable to ensure transfusion safety is extremely valuable [8]. In addition, the disinfection procedure must offer minimal risk to host cells [7, 8].

Special attention should be given about photosensitizers (PS) that can be used for blood disinfection. Due to the heterogeneous nature of blood tissue including all its cellular and soluble components, photoinactivation must be exclusively directed to pathogens. It is noteworthy that the PS and its photoproducts must be nontoxic and nonimmunogenic to the acceptor [9]. In this context, methylene blue (MB) and crystal violet or gentian violet (CV) are the most tested PS in blood decontamination procedures. Whereas these PS compounds have a history of success against several human pathogens present in transfusion blood and its products, there are only few studies in veterinary medicine, and therefore, further investigations should be encouraged.

Over the past 40 years, MB-PDT has been extensively studied showing that it can inactivate all sorts of human pathogens, i.e., bacteria, fungi, protozoa, and viruses [7, 10–15]. MB is produced by several pharmaceutical companies and is cost-effective and absorbs long-wavelength light (660 nm) that favors light transmission in blood since hemoglobin poorly absorbs this wavelength. Due to all these positive features, European transfusion centers routinely use MB to decontaminate blood plasma since 1992 [7, 8]. However, MB-PDT is not suitable to decontaminate whole cell blood because it causes extensive hemolysis due to membrane damage and protein denaturation within cells [7, 9]. Another limitation of PDT-treated plasma occurs due to degradation of factor VIII and fibrinogen [8] inhibiting coagulation function.

CV was used for many years to inhibit Chagas' disease (caused by protozoan *Trypanosoma cruzi*) transmission via blood transfusion but was discontinued in the beginning of this decade due to side effects [9]. It turns out that CV antiparasitary activity was being achieved by intrinsic toxicity of CV when present in high concentrations, but not due to photodynamic reactions. To avoid toxicity caused by high-CV concentration and actually use its photodynamic potential, Ramirez and colleagues [16] performed experiments to evaluate photoinactivation of *T. cruzi* by combining ascorbic acid (vitamin C) with low-CV concentrations and visible light. Ascorbic acid is a well-known hydrophilic antioxidant that can protect soluble proteins against oxidative damage. Interestingly, ascorbic acid significantly enhances the killing efficiency of CV-PDT activity against *T. cruzi* in blood, when compared to CV-PDT alone [10, 17, 18]. Moreover, the concentration of CV used was significantly lower than previous recommendations, allowing further minimization of side effects [16].

13.4 Disinfection of Water and Prevention of Waterborne Diseases

Water is considered a renewable natural resource, but despite the means of natural recycling, its use in animal production and agriculture faces a major concern among international organizations and governments. Concerns arise not only from increasing prices but also because of losses of natural reservoirs due to dumping of untreated residues. Massive administration of potent antimicrobial agents to animal production is adopted worldwide to prevent epidemics in farms and to accelerate weight gain. It is surely a comprehensive strategy to decrease the pool of infected patients. However, this is a high-risk procedure when we consider environmental risk of exposure to antimicrobial drugs. As a source of evolutive selection pressure, this attitude facilitates the development and dissemination of drug-resistant pathogens [19, 20]. Therefore, production farm residues have already become a major issue that needs to be addressed in a consistent way.

PDT has been recently cited as potential technology for the disinfection of polluted water ponds of fish-farming and also to treat infections in fish and crustaceans [21–25]. Conversely, PDT efficiency to decontaminate drinking water and wastewater was proved by laboratory tests of fecal coliforms detection [22, 26–30]. The possibility to use sunlight as a natural and free source of light to treat water makes this technology attractive in relation to its cost-effectiveness [22]. The assumption that PDT could be an effective strategy to reduce the use of antibiotics in shrimp hatchery systems was studied by Asok et al. [31]. Their experiment reported successful use of rose bengal and white light to inactivate drug-resistant *Vibrio harveyi*. In conclusion, they propose that PDT could be suitable to reduce the use of antibiotics in shrimp larviculture systems and avoid hazards related to human health and the ecosystem.

Fish parasites are among the most important infectious diseases that could heavily damage fish aquaculture. Massive losses occur due to high mortality and morbidity rates of adult animals and loss of fish eggs and young larvae [32, 33]. An effective substance, malachite green, was successfully employed against the majority of fish parasites, but because of carcinogenic potential, it is no longer permitted to treat fish for human consumption [32–34]. As alternative, other substances were tested, such as sodium chloride, peroxide, formaldehyde, cooper sulfate, and potassium permanganate, but they are prohibitively expensive and did not exhibit very effective results [33]. On the other hand, it was shown that porphyrins, when used as PS, do not exhibit significant toxicity for fish [21]. According to Rotomskis et al. and Alves et al., another major advantage of this technique is that excessive accumulation of porphyrins in the environment seems very unlikely due to its posterior degradation under the action of sunlight [22, 29]. Magaraggia et al. investigated the photodynamic activity of porphyrin derivatives activated by white light against waterborne pathogens. The authors evaluated two different cationic porphyrins against methicillin-resistant *Staphylococcus aureus*, *Escherichia coli*, and *Saprolegnia* spp. strains, as a model of contaminated aquaculture water. They also studied the photodynamic treatment of spontaneously and artificially *Saprolegnia*-infected rainbow trout (*Oncorhynchus mykiss*) as a model for clinical treatment of saprolegniasis. The results showed a significant reduction of bacterial and fungal populations after short-irradiation procedures, leading to successful water disinfection and cure of infected animals [21]. In 2011, Arrojado et al. also evaluated the capacity of PDT to inactivate nine species of fish pathogenic bacteria often associated with aquaculture system contamination (e.g., *Vibrio anguillarum*, *Vibrio parahaemolyticus*, *Aeromonas salmonicida*, *Photobacterium damselae* subsp. *damselae*, *Photobacterium damselae* subsp. *piscicida*, *E. coli*, *Pseudomonas* spp., *Enterococcus faecalis*, and *S. aureus*). All tested strains could be inactivated by a tricationic porphyrin derivative excited by white light [30].

Ichthyophthiriasis, known as white spot disease, occurs in many freshwater fish species due to its low-host specificity and is caused by the protozoan *Ichthyophthirius multifiliis*. In vitro and in vivo studies, respectively, performed by Wohllebe et al. and Häder et al. have tested the photodynamic activity of chlorophyllin, a nontoxic water-soluble chlorophyll derivative, to kill different life stages of the protozoan [32, 33]. The same light source and irradiance were also used: a system with broad-spectrum white light, which simulates solar emission (i.e., 400–700 nm, 149.66 mW/cm^2; UV-A, 32.67 mW/cm^2; UV-B 0.77 mW/cm^2). All experiments obtained promising results against the protozoans, reducing number and multiplication capacity of the parasites. Following their noteworthy results, Häder et al. attempted to kill other species of fish parasites. The authors employed the same PDT protocol for in vitro assays against wild strains of *Ichthyobodo necator*, *Dactylogyrus* spp., *Trichodina* spp., and *Argulus* spp. Once again, PDT was effective against all parasites except for the crustacean *Argulus* spp. [34]. Thus, based on all abovementioned studies, we propose PDT as a valid, eco-friendly, and feasible method to be employed against several microbial contaminants in ponds and aquaculture.

13.5 Decontamination of Surfaces

The evolutive selection of bacterial strains resistant to virtually all commercially available antibiotics has emerged as one of the great challenges of medicine in the twenty-first century. A major concern on dissemination of drug resistance and other virulence-related genes relies on the bacterial ability to conjugate plasmids. Due to the massive use of antimicrobials combined to the presence of highly contaminated individuals and waste, hospitals and production farms currently represent the most fertile environments for microbial drug-resistant selection and dissemination.

Therefore, insufficient surface decontamination in such environments can lead to cross-contamination among individuals and, ultimately, the occurrence of epidemics within the facilities.

Current methods for surface and air decontamination based on the use of chemicals and gases are indeed effective for many situations, even though they are obviously not free of limitations. Problematic examples can be related to product or material damage, development of microbial resistance, and exposure of people and animals to potentially harmful residues. Yet, pathogens such as *Mycobacteria* are intrinsically tolerant to chemical treatments due to their protective cell envelope that restricts the uptake of many drugs and disinfectant chemicals. In order to overcome the challenges of chemical decontamination of surfaces, the adoption of additional or alternative approaches must be applied.

The use of UVC light may be addressed as highly effective strategy. However, the limitations are still many: low penetration even into transparent materials, such as glass and plastics; degradation of products and materials; and its ability to provide harm to animal health due to its direct damage or the photochemical formation of ozone. As demonstrated by the Nobel Prize worthy observation made by Niels Finsen over a century ago, blue light does not induce deleterious or inflammatory effects in animals, as did UV radiation, but could still kill bacteria including those from *Mycobacterium* gender (further details in Chap. 1). At the time, Finsen was not aware of this fact, but several microbial species are naturally sensitive to blue light because they accumulate intracellular stocks of photodynamically active porphyrins such as uroporphyrin, coproporphyrin, and protoporphyrin. The inactivation kinetics presented among different species and strains is dependent on porphyrin concentration levels, porphyrin types, expression of antioxidant defense, and wavelength of excitation [35].

With the advent of high-power LED technologies, dedicated devices have been developed under optimized conditions to inactivate a broad spectrum of microorganisms using high-intensity narrow-spectrum (HINS) light. This technology platform represents a much safer and versatile approach when compared to UVC irradiation. To illustrate its safety level, the same strategy has been used over the past years to treat acne lesions leaving no side effects. Supporting data published by Dai et al. yet indicates that *P. aeruginosa* is much more susceptible to blue light inactivation than mammalian cells using well-controlled in vitro and in vivo assays [35]. As a result, it can be assumed that there is a wide therapeutic window where bacteria can be selectively inactivated by blue light, while the host tissue cells are preserved. For surface decontamination, many studies have already demonstrated positive results of HINS light against a broad spectrum of nosocomial and food pathogens [36, 37]. Other important advantage of HINS irradiation is its capacity to effectively inactivate microorganisms in the naturally resistant forms such as spores and biofilms, which are very commonly present in contaminated surfaces [38, 39]. Biofilms can be up to 1,000-fold more resistant to chemical antimicrobials, including exogenous PS, when compared to cells in planktonic suspension. This characteristic is associated to several biofilm features: impermeability of extracellular matrix layer, gradient of sub-minimum inhibitory concentration inside the biofilm, alteration of metabolism, horizontal genetic transfer, elevated concentration of antibiotic degrading enzymes, etc. However, these resistance mechanisms impose none to minor barriers for HINS since it does not depend on drug administration; thus, limiting factors are basically light irradiance and concentrations of oxygen and endogenous photosensitizers. Microbial spores are also famous in respect to its resistance to intense physical and chemical stress. Exposures to heat, ionizing (gamma and ultraviolet) radiation, desiccation, iodine, and alkylating agents are a few examples of disinfectants that spores can tolerate. For photoinactivation it is not very different. According to Maclean et al., endospores of *Bacillus cereus* and *Clostridium difficile* require tenfold higher HINS exposure time to be inactivated when compared to vegetative phase cells [40]. However, since HINS light offers no harm to us and current LED technology is very energy efficient, it is feasible to have a device working full time over the most critical surfaces. The only evident disadvantage of HINS decontamination is its inactivity against viral particles since they do not accumulate cellular metabolites such as derivatives of porphyrins or flavins [41]. To facilitate comparison between UV and 405 nm light, we reproduced in Table 13.1 a summarized list of their most relevant characteristics.

Table 13.1 Comparison of the properties of ultraviolet C (UVC) and 405 nm light for environmental disinfection applications

	UVC light	405 nm light
Typical/potential use	Terminal clean of air- and light-exposed surfaces	Continuous disinfection of air- and light-exposed surfaces
Safety	Significant safety hazards associated with human exposure can cause DNA and tissue damage	Can be used safely in the presence of people at recommended irradiance levels
Mechanism of action	Indiscriminate DNA and protein damage. Sublethally damaged cells can repair DNA (e.g., UVR-system, photolyase enzyme)	Photo-excitation of intracellular molecules induces oxidation of microbial cells
Antimicrobial activity	Broad-spectrum action against microorganisms including spores and viruses	Effective against bacteria, fungi, yeasts, and spores; antiviral activity not yet established
Antimicrobial efficacy	Rapid inactivation rate within treatment zone. Normally within less than 60 min	Comparably slower inactivation rate within treatment zone. Normally within less than 3 h
Materials compatibility	UV-light-associated polymer damage	Lower energy of 405 nm photons do not damage polymers
Ease of use for environmental disinfection	Rooms/wards need to be vacated during use; operator training required	Can be safely used during room occupation; no operator safety training required
Microbial mutagenic potential	Powerful mutagen	Multi-target oxidative action mitigates against resistance development
Penetrability	Does not penetrate through plastics or glass and weakly penetrates into water and fabrics. Very low penetration into biologic tissue	Can penetrate through plastics, glass, water, and fabrics. Low penetration into biologic tissue

Reproduced from Maclean et al. [41]

References

1. Tezel TH, Bora NS, Kaplan H. Pathogenesis of age-related macular degeneration. Trends Mol Med. 2004;10(9):417–20.
2. Olivo M, Bhuvaneswari R, Lucky SS, Dendukuri N, Soo-Ping Thong P. Targeted therapy of cancer using photodynamic therapy in combination with multi-faceted anti-tumor modalities. Pharmaceuticals. 2010;3(5):1507–29.
3. Kawczyk-Krupka A, Bugaj AM, Potempa M, Wasilewska K, Latos W, Sieroń A. Vascular-targeted photodynamic therapy in the treatment of neovascular age-related macular degeneration: clinical perspectives. Photodiagnosis Photodyn Ther. 2015;12(2):161–75.
4. Gryson O. Servier medical art France: servier. 2016. Available from: http://www.servier.com/Powerpoint-image-bank.
5. Cooper AE, Ahonen S, Rowlan JS, Duncan A, Seppälä EH, Vanhapelto P, et al. A novel form of progressive retinal atrophy in Swedish vallhund dogs. PLoS ONE. 2014;9(9):e106610.
6. Petersen-Jones SM, Komáromy AM. Dog models for blinding inherited retinal dystrophies. Hum Gene Ther Clin Dev. 2015;26(1):15–26.
7. Aureliano DP, Ribeiro MS, Lindoso JAL, Pogliani FC, Sellera FP, Song D, et al. Treatment and control of leishmaniasis using photodynamic therapy. In: Leishmaniasis – trends in epidemiology, diagnosis and treatment. InTech; Rijeka, Croacia; 2014.
8. Wainwright M. Pathogen inactivation in blood products. Curr Med Chem. 2002;9(1):127–43.
9. Wainwright M, Baptista MS. The application of photosensitisers to tropical pathogens in the blood supply. Photodiagnosis Photodyn Ther. 2011;8(3):240–8.
10. Wilson M, Dobson J, Harvey W. Sensitization of oral bacteria to killing by low-power laser radiation. Curr Microbiol. 1992;25(2):77–81.
11. Schagen FH, Moor AC, Cheong SC, Cramer SJ, van Ormondt H, van der Eb AJ, et al. Photodynamic treatment of adenoviral vectors with visible light: an easy and convenient method for viral inactivation. Gene Ther. 1999;6(5):873–81.
12. Zeina B, Greenman J, Purcell WM, Das B. Killing of cutaneous microbial species by photodynamic therapy. Br J Dermatol. 2001;144(2):274–8.
13. Zolfaghari PS, Packer S, Singer M, Nair SP, Bennett J, Street C, et al. In vivo killing of Staphylococcus aureus using a light-activated antimicrobial agent. BMC Microbiol. 2009;9(1):27.
14. Song D, Lindoso JAL, Oyafuso LK, Kanashiro EHY, Cardoso JL, Uchoa AF, et al. Photodynamic therapy using methylene blue to treat cutaneous leishmaniasis. Photomed Laser Surg. 2011;29(10):711–5.
15. Gilaberte Y, Aspiroz C, Alejandre MC, Andres-Ciriano E, Fortuño B, Charlez L, et al. Cutaneous sporotrichosis treated with photodynamic therapy: an in vitro and in vivo study. Photomed Laser Surg. 2014;32(1):54–7.
16. Ramirez LE, Lages-Silva E, Pianetti GM, Rabelo RM, Bordin JO, Moraes-Souza H. Prevention of transfusion-associated Chagas' disease by sterilization of Trypanosoma cruzi-infected blood with gentian violet, ascorbic acid, and light. Transfusion. 1995;35(3):226–30.
17. Docampo R, Moreno SN, Cruz FS. Enhancement of the cytotoxicity of crystal violet against Trypanosoma cruzi in the blood by ascorbate. Mol Biochem Parasitol. 1988;27(2–3):241–7.
18. Gadelha FR, Hanna PM, Mason RP, Docampo R. Evidence for free radical formation during horseradish peroxidase-catalyzed N-demethylation of crystal violet. Chem Biol Interact. 1992;85(1):35–48.
19. Dahshan H, Abd-Elall AMM, Megahed AM, Abd-El-Kader MA, Nabawy EE. Veterinary antibiotic resistance, residues, and ecological risks in environmental samples obtained from poultry farms. Egypt Environ Monit Assess. 2015;187(2):2.
20. Venglovsky J, Sasakova N, Placha I. Pathogens and antibiotic residues in animal manures and hygienic and ecological risks related to subsequent land application. Bioresour Technol. 2009;100(22):5386–91.

21. Magaraggia M, Faccenda F, Gandolfi A, Jori G. Treatment of microbiologically polluted aquaculture waters by a novel photochemical technique of potentially low environmental impact. J Environ Monit. 2006;8(9):923–31.
22. Alves E, Faustino MAF, Tomé JPC, Neves MGPMS, Tomé AC, Cavaleiro JAS, et al. Photodynamic antimicrobial chemotherapy in aquaculture: photoinactivation studies of Vibrio fischeri. PLoS ONE. 2011;6(6):e20970.
23. Jori G, Magaraggia M, Fabris C, Soncin M, Camerin M, Tallandini L, et al. Photodynamic inactivation of microbial pathogens: disinfection of water and prevention of water-borne diseases. J Environ Pathol Toxicol Oncol. 2011;30(3):261–71.
24. Coppellotti O, Fabris C, Soncin M, Magaraggia M, Camerin M, Jori G, et al. Porphyrin photosensitised processes in the prevention and treatment of water- and vector-borne diseases. Curr Med Chem. 2012;19(6):808–19.
25. Fabris C, Soncin M, Jori G, Habluetzel A, Lucantoni L, Sawadogo S, et al. Effects of a new photoactivatable cationic porphyrin on ciliated protozoa and branchiopod crustaceans, potential components of freshwater ecosystems polluted by pathogenic agents and their vectors. Photochem Photobiol Sci. 2012;11(2):294–301.
26. Jemli M, Alouini Z, Sabbahi S, Gueddari M. Destruction of fecal bacteria in wastewater by three photosensitizers. J Environ Monit. 2002;4(4):511–6.
27. Bonnett R, Krysteva MA, Lalov IG, Artarsky SV. Water disinfection using photosensitizers immobilized on chitosan. Water Res. 2006;40(6):1269–75.
28. Carvalho CMB, Tomé JPC, Faustino MAF, Neves MGPMS, Tomé AC, Cavaleiro JAS, et al. Antimicrobial photodynamic activity of porphyrin derivatives: potential application on medical and water disinfection. J Porphyrins Phthalocyanines. 2009;13(04n05):574–7.
29. Rotomskis R, Streckyte G, Bagdonas S. Phototransformations of sensitizers 2. Photoproducts formed in aqueous solutions of porphyrins. J Photochem Photobiol B Biol. 1997;39(2):172–5.
30. Arrojado C, Pereira C, Tomé JPC, Faustino MAF, Neves MGPMS, Tomé AC, et al. Applicability of photodynamic antimicrobial chemotherapy as an alternative to inactivate fish pathogenic bacteria in aquaculture systems. Photochem Photobiol Sci. 2011;10(10):1691–700.
31. Asok A, Arshad E, Jasmin C, Pai SS, Singh ISB, Mohandas A, et al. Reducing Vibrio load in Artemia nauplii using antimicrobial photodynamic therapy: a promising strategy to reduce antibiotic application in shrimp larviculture. Microb Biotechnol. 2012;5(1):59–68.
32. Wohllebe S, Richter P, Häder D-P. Chlorophyllin for the control of ichthyophthirius multifiliis (fouquet). Parasitol Res. 2012;111(2):729–33.
33. Häder D-P, Schmidl J, Hilbig R, Oberle M, Wedekind H, Richter PR. Treatment of ichthyophthiriasis with photodynamically active chlorophyllin. Parasitol Res. 2016;115(4):1509–17.
34. Häder D-P, Schmidl J, Hilbig R, Oberle M, Wedekind H, Richter P. Fighting fish parasites with photodynamically active chlorophyllin. Parasitol Res. 2016;115(6):2277–83.
35. Dai T, Gupta A, Huang Y-Y, Yin R, Murray CK, Vrahas MS, et al. Blue light rescues mice from potentially fatal Pseudomonas aeruginosa burn infection: efficacy, safety, and mechanism of action. Antimicrob Agents Chemother. 2013;57(3):1238–45.
36. Maclean M, MacGregor SJ, Anderson JG, Woolsey G. Inactivation of bacterial pathogens following exposure to light from a 405-nanometer light-emitting diode array. Appl Environ Microbiol. 2009;75(7):1932–7.
37. Murdoch LE, Maclean M, Endarko E, MacGregor SJ, Anderson JG. Bactericidal effects of 405 nm light exposure demonstrated by inactivation of Escherichia, Salmonella, Shigella, Listeria, and Mycobacterium species in liquid suspensions and on exposed surfaces. Sci World J. 2012;2012:1–8.
38. St Denis TG, Dai T, Hamblin MR. Killing bacterial spores with blue light: when innate resistance meets the power of light. Photochem Photobiol. 2013;89(1):2–4.

39. Wang Y, Wu X, Chen J, Amin R, Lu M, Bhayana B, et al. Antimicrobial blue light inactivation of gram-negative pathogens in biofilms: in vitro and in vivo studies. J Infect Dis. 2016;213(9):1380–7.
40. Maclean M, Murdoch LE, MacGregor SJ, Anderson JG. Sporicidal effects of high-intensity 405 nm visible light on endospore-forming bacteria. Photochem Photobiol. 2013;89(1):120–6.
41. Maclean M, McKenzie K, Anderson JG, Gettinby G, MacGregor SJ. 405 nm light technology for the inactivation of pathogens and its potential role for environmental disinfection and infection control. J Hosp Infect. 2014;88(1):1–11.

Chapter 14
Future Perspectives

Fábio Parra Sellera, Cristiane Lassálvia Nascimento, Fabio Celidonio Pogliani, Caetano Padial Sabino, and Martha Simões Ribeiro

Abstract Nowadays, it is clear that the activity of different photosensitizers (PSs) has a strong potential for moving photodynamic therapy (PDT) to clinical practice. Present technologies as dedicated light sources, new PSs, and nanotechnology are emerging strategies to promote PDT as a reliable, cost-effective, and safe approach to veterinary medicine. This chapter addresses an overview of emerging clinical applications and recent technologies to encourage veterinarians toward PDT.

F.P. Sellera (✉) • F.C. Pogliani
Department of Internal Medicine, School of Veterinary Medicine and Animal Science, University of São Paulo, Av. Prof. Orlando Marquês Paiva, 87, Butanta 13635-900, SP, Brazil
e-mail: fsellera@usp.br; fabiocp@usp.br

C.L. Nascimento
Veterinary Unit of Santos Aquarium, Santos Aquarium, Av. Bartolomeu de Gusmão S/N, Ponta da Praia, Santos 11035-500, SP, Brazil
e-mail: cristianelassalvia@hotmail.com

C.P. Sabino
Department of Microbiology, Institute for Biomedical Sciences, University of São Paulo, Av. Lineu Prestes 1347, Cidade Universitária, Sao Paulo 05508-000, SP, Brazil

Department of Clinical Analysis, School of Pharmaceutical Sciences, University of São Paulo, Sao Paulo, SP, Brazil

Center for Lasers and Applications, Nuclear and Energy Research Institute, National Commission for Nuclear Energy, Av. Lineu Prestes 2242, Cidade Universitária, Sao Paulo 05508-000, SP, Brazil
e-mail: caetanosabino@gmail.com

Department of Medical Biophysics, Princess Margaret Cancer Institute/University of Toronto, Toronto, Canada

M.S. Ribeiro
Center for Lasers and Applications, Nuclear and Energy Research Institute, National Commission for Nuclear Energy, Av. Lineu Prestes 2242, Cidade Universitária, Sao Paulo 05508-000, SP, Brazil
e-mail: marthasr@usp.br

F.P. Sellera et al. (eds.), *Photodynamic Therapy in Veterinary Medicine: From Basics to Clinical Practice*, DOI 10.1007/978-3-319-45007-0_14

14.1 Introduction

In the previous chapters, we presented some aspects related to photodynamic therapy (PDT) as its history, mechanisms, and applications that deserve special attention. In recent decades, the number of researches on PDT in veterinary medicine has increased. New photosensitizers (PSs) or functionalized PSs are being developed to optimize PDT (see Chap. 3) as well as dedicated light sources. In fact, innovative light sources for light-based therapies have been widely explored. Recently, Chinese researchers have developed a "3-D bright fabric," composed of flexible polymer optical fibers and LED at red emission to be used on the human body injuries in the treatment of diseases of various origins [1].

Several studies have demonstrated a wide perspective in clinical treatment of diseases, such as topical infections and cancer. Moreover, environmental applications, as water treatment in fish farming, are promising candidates to reduce antimicrobial residues. As we mentioned above, the number of veterinary PDT studies has increased; on the other hand, veterinary medicine still has difficulties to determine clinical protocols due to a wide variety of animal species with different characteristics and particular diseases. Many well-established studies in other areas indicate applications not yet explored in veterinary medicine. In this chapter, we will discuss some relevant aspects regarding worthy PDT as new frontiers of research.

14.2 Prospects for Clinical Applications

Technological advances in recent decades have improved the diagnosis and treatment of diseases that used to be hardly managed in the past. In previous chapters, we presented a series of PDT studies in veterinary medicine that proved that this technique could be widely investigated in this field. Besides that, the growing concern about the overuse of antibiotics, resulting in the appearance of multidrug-resistant microorganisms, such as the absence of a 100 % effective treatment against the various types of cancer, makes PDT a promising candidate for veterinarians.

As previously discussed elsewhere, antineoplastic PDT has been more studied than antimicrobial PDT in dogs and cats. A lack on antimicrobial therapies is not an excuse to slow enhancements in cancer studies which should be more explored by veterinarians. New alternatives are being established by the development of new photosensitizers and irradiation systems that could help on different types of cutaneous tumors treatments, becoming new strategies against this illness. Studies already developed in human medicine stimulate employment in routine dermatological practice, which certainly can be extrapolated for veterinary medicine in the near future.

14.2.1 Skin Diseases

Skin infections are commonly diagnosed in companion animals. Thus, PDT could benefit Veterinary Dermatology. The direct and easy approach, with few physical barriers, makes its application more suitable than for internal organs, given that they are more difficult to be irradiated and accessible for dyes. Furthermore, studies on pathogens from human diseases similar to those that affect animals have proved to be susceptible to PDT, e.g., *Staphylococcus aureus*, *Staphylococcus epidermidis*, *Staphylococcus intermedius*, *Pseudomonas aeruginosa*, *Escherichia coli*, *Proteus* spp., *Microsporum* spp., *Trichophyton* spp., *Sporothrix* spp., and *Malassezia* spp. These results motivate the development of clinical trials focusing attention on topical treatments and targeting antimicrobials to systemic therapies.

Localized and integumentary infections and skin cancer are among eligible diseases that PDT can be useful for farm animals. Swapping the usual systemic for topical treatment fulfills the reasons to develop future researches. For instance, habronemiasis (*Habronema* spp.) and pythiosis (*Pythium insidiosum*), a helminth and an oomycete agent, respectively, are severe equine skin infectious diseases with relative high occurrence in tropical countries. In addition, equine sarcoid is the most commonly diagnosed tumor of the skin and soft tissues. These diseases combine surgical and systemic drugs for treatment, but the rate of unsuccessful results is high leading to euthanasia. PDT could be tested as substitute or adjuvant therapy.

Udder and ear helminthiases in cows, caused by *Stephanofilaria* spp. and *Rhabditis* spp., respectively, still affect many animals worldwide. Protocols of residue control avoid antimicrobial systemic drugs during milk and meat withdrawal periods. In many cases, topical drugs are ineffective. Thus, PDT could be employed as an alternative or combined therapy with topical agents.

Besides the ability to treat infections and cancer, there is indication of PDT to treat macular degeneration in humans, as commented in Chap. 13. Animals also can be affected by disorders involving neovascularization. Lucroy raised a very interesting question not yet investigated [2]. In fact, he suggested that PDT could be applied to combat formation of exuberant granulation tissue in horses. Horses have specific dermal and subcutaneous precursors that predispose to a local exuberant granulation tissue preventing the wound contraction and reepithelialization. A similar principle of selectively destroying functional blood vessels, as described for macular degeneration in humans, could be investigated for exuberant granulation tissue in horses, using appropriate PSs as those reported in Chap. 3.

The wild animal medicine seems to open a huge range to develop researches involving PDT, once the knowledge about different wild species, metabolism, diseases, and their treatments is scarce.

Successful treatment of footpad dermatitis in penguins encourages us to expand the list of species, which suffer from similar diseases as is the case of birds of prey, waders, Galliformes, among others. Similarly, dermatitis caused by poor captive conditions that affects many species, such as bedsores due to contact with urine and feces, moisture, and imprisonment in restricted areas, can be benefitted from this broad-spectrum treatment against bacteria and fungi.

Until now, there are no reports of PDT in amphibians. Thus, we believe that challenging conditions, such as the chytridiomycosis (*Batrachochytrium dendrobatidis*), which threaten to extinguish several species of amphibians, deserve to be studied as PDT application viability.

14.2.2 Oral Diseases

Veterinary dentistry is another potential area for PDT application. PDT has been widely studied in human dentistry to treat dental, mucosal infections and cancer presenting benefits in clinical practice.

Although dogs have been used as experimental model, PDT was successful to treat induced periodontitis and periimplantitis (see Chap. 12). Nevertheless, veterinary medicine still has not explored this practice, and other diseases could be also treated based on human studies.

An interesting PDT approach may be to treat epulis. Epulis is considered the most common tumor in dog's mouth. In severe cases, it could lead to fracture due to bone involvement. Although histologically benign and with favorable prognosis, epulis must be surgically removed, and in some cases the association with radiotherapy is recommended. Recently, Truschnegg et al. reported an epulis case in human treated by methylene blue (MB)-mediated PDT and red laser [3]. No sign of recurrence of any hyperplastic tissue was observed after 4-week follow-up and even after 12 months. These results are motivating to extrapolate this condition to veterinary's use.

Stomatitis in reptiles is another interesting approach to be addressed in the near future. Our preliminary results (see Chap. 12) suggest that PDT was effective to treat this disease in snakes. Birds are also affected by oral infections, including oral and beak infections. According to our previous experience, we encourage further studies to perform PDT for this purpose.

14.2.3 Diseases Related to Other Organs

Nowadays, there has been growing public concern related to the persistence of drug residues in milk products and their consumption by humans. Diseases like mastitis are the main concern worldwide once the same drugs could be used by human medicine. It is well known that continued or indiscriminate use of antibiotics for mastitis treatment has been favoring the increase of multiresistant microorganisms (MOs) and limiting effective strategies to eradicate this problem in milk farms. New PDT appliances can be designed in order to overwhelm those challenges, and the antibiotic use becomes strict to indispensable medical cases.

Mastitis is one of the most prevalent diseases in dairy cattle. Production losses and early culling rates due unsuccessfully treatments are negative effects related to economic losses. Because bacterial infections are considered the most prevalent cause, antimicrobial therapy is always required.

An unusual type of mastitis caused by algae *Prototheca zopfii*, also a zoonosis, remains with no effective treatment and cure [4]. Likewise the mastitis caused by *S. aureus*, clinical presentation is severe and difficult to control. It is also potentially infectious to other animals in the herd. For both *P. zopfii* and *S. aureus*, the recommendation is culling the infected animals. Recently, we investigated PDT in vitro against different mastitis pathogens, i.e., bacteria and *P. zopfii*. The results were successful (see Chap. 11), but more realistic experiments as in milk samples have to be conducted before clinical trials [5].

PDT is also well renowned in cancer treatment to treat dysplastic lesions and malignant types of cancer by endoscopy. In these situations, diffusing fibers are especially developed to irradiate cavities such as bladder, esophagus, lung, and stomach. At this point, PDT could emerge as an alternative to treat different types of internal cancer. This approach has not been explored, but there is a great perspective for its use.

Another interesting issue to be investigated is the use of PDT to treat osteomyelitis in dogs and cats. Some studies conducted by Tardivo et al. presented successful results in human patients [6, 7]. In those studies, PDT was able to prevent foot amputation in diabetic patients. As is well known, various traumas (e.g., running over) may lead to extensive damage of the skin providing a gateway to infection and contributing to MO proliferation in osteomuscular tissues. Osteomyelitis in companion animals, as in any other species, requires prolonged treatment based on cocktail drugs including anti-inflammatories and systemic antibiotics [8]. Often these drugs are not able to control the infection eventually leading to amputation or euthanasia [9]. In this context, PDT may emerge, if not to replace, in association with other clinical protocols routinely used.

Bacterial and fungal ulcerative keratitis, as well as non-ulcerative fungal keratitis as stromal abscess, is frequent in horses. Environmental and behavioral factors are the main cause to make horses more susceptible to corneal and conjunctival lesions than other domestic animals, since these structures are frequently exposed to bacteria and fungi, especially *Aspergillus* spp. and *Fusarium* spp. Recent study evaluated clinical outcome in equine keratomycosis reporting that surgical intervention was necessary for 54 % of the eyes, and 28 % of these eyes were enucleated [10].

PDT could be an alternative approach for those cases. Since the 1970s of the last century, researchers use animal models to test PDT in virus-induced keratitis with good outcome [11]. More recently, Shih and Huang used MB-mediated PDT combined with amikacin to treat nontuberculous mycobacterial keratitis in rabbits [12]. The authors indicated PDT as a potential adjuvant treatment for intractable mycobacterial infection. In 2015, Zborovska and Dorokhova presented in the 15th EURETINA Congress, held in Nice, France, data about PDT on fungal inflammatory eye diseases from studies in vitro to human clinical practice. In vitro assays determined the most effective parameters of MB-mediated PDT. Preclinical studies in rabbits revealed that control group, which received standard anti-inflammatory (AI) therapy, had disease duration higher than PDT + AI group (about 7 days for moderate and severe keratitis). Clinical outcome showed that after 3 months, a proportion of patients with corneal infiltrate and erosion area in PDT group was lower than that of control group.

14.2.4 Environmental Applications

Phototoxicity against pathogens could be enhanced by using sunlight (see Chap. 13). The idea of using compounds activated by sunlight is old, but seems to be an interesting option to control parasites and fungal diseases, mainly in water. In fact, Chap. 13 addressed recent works regarding this issue with encouraging results.

PDT could be investigated to inactivate other parasitic agents as monogenetic trematodes (gill fluke, i.e., *Dactylogyrus* spp., *Gyrodactylus* spp., *Cleidodiscus* spp.), microcrustaceans (*Argulus* spp., *Ergasilus* spp., *Lernaea* spp.), cestodes, and nematodes.

Other fish parasitic diseases could be also explored by PDT as marine (*Cryptocaryon irritans*) or freshwater protozoa (*Ichthyophthirius multifiliis*, *Epistylis* spp., *Ambiphrya* spp., *Trichodina* spp., and *Trichophrya* spp.). Fungi (*Saprolegnia parasitica* on aquatic invertebrates) and bacteria (*Aeromonas hydrophila*, *Pseudomonas fluorescens*, *Flexibacter columnaris*, *Streptococcus* spp.) may be sensitive to PDT as seen in Chap. 12. The main challenge is the technical adaptation to the aquatic environment.

Similarly, sunlight-based PDT appears to be interesting alternative to reduce parasites on pasture. The idea is based on the oral supplementation of animals with nontoxic photosensitizers to break parasite cycle. When the animal eliminates parasites, the sun can inactivate them. The published studies regarding this topic were detailed in Chap. 12.

14.2.5 Biotechnology Applied to Animal Reproduction

As we discussed in Chap. 11, another application that can be envisaged for PDT involves biotechnology applied to animal reproduction. Due to many ethical questions, the investigation of PDT on human reproductive system to treat embryos and semen infections seems to be more limited than veterinary. By the other hand, future researches involving animals could promote an improvement of knowledge and safety, which could be extrapolated for humans.

In poultry, semen samples are usually contaminated with feces and urine. PDT can be performed to reduce microbial load improving the sperm motility and fertilization rate. Furthermore, this application could be extended to conservation of endangered species, improving animal fertility rate in captivity and, consequently, increasing the success rates in their conservation.

Endemic diseases in cattle are spread worldwide. Thus, artificial insemination could be a high-risk potential vector to infect animals due to the large-scale commerce of cattle semen among countries. Following international recommendations provided by the World Organization for Animal Health (OIE) to evaluate the health status of bulls in artificial insemination stations, several tests are being made in animals and their semen. However, diseases with slow transmission are more difficult to monitor due to their seroconversion [13], and diagnostic tests on semen present some limitations [14, 15]. Therefore, the risk of pathogen transmission in artificial insemination procedures remains and challenges new practices for its containment.

Addition of antibiotics in the media is a satisfactory safety for bacterial diseases, mainly campylobacteriosis [13]. However, other pathogens could lead to infections due to the inefficacy of antimicrobials against them. The main relevant diseases that could be transmitted by semen are viral (foot-and-mouth disease, enzootic bovine leukosis, infectious bovine rhinotracheitis, infectious pustular vulvovaginitis, rinderpest, bluetongue, bovine diarrhea, malignant catarrhal fever, Akabane virus), bacterial (contagious bovine pleuropneumonia, Johne's disease, brucellosis, bovine tuberculosis, leptospirosis, bovine genital campylobacteriosis, Query fever (hemorrhagic septicemia)), protozoan (bovine genital trichomoniasis), and even prion (bovine spongiform encephalopathy). The PDT inactivation of these agents should be evaluated in raw and collected semen from bulls.

In addition to the semen approach, the disinfection of embryos in farm animals also appears to be an attractive strategy to control transmission of reproductive diseases. Embryo technologies are constantly evolving; however, it is imperative to contain the risk factors and assure the health status of the farm, herd, donor cow, and embryo to achieve full success in livestock reproduction programs. Thus, the risk assessment of potential pathogens, preventive measures, diagnosis, and disease control has become a true challenge nowadays.

Sutmoller and coworkers made simulations to evaluate the risks of transmission of viral agents by embryos after following the recommendations for international trade based on epidemiology and surveillance as well as the internationally approved embryo processing protocols [16]. The authors concluded that the foot-and-mouth disease virus, vesicular stomatitis virus, and bluetongue virus are with a very low risk. In another research done by Wrathall and colleagues, they studied embryos produced with infected semen for enzootic bovine leukosis, bovine herpesvirus-1, bovine viral diarrhea virus, and bluetongue virus [17]. After embryo processing, they concluded that even with an extremely low health risk, virus removal from these embryos is difficult with the exception of enzootic bovine leukosis.

The majority of the studies focus on viral infections because viruses represent a great risk of infection of the embryo due to their too small size [18]. Therefore, more studies are necessary to evaluate different types of MOs that could be transmitted by reproduction techniques and probably be prevented by PDT.

Embryo transfer by in vivo and in vitro techniques is used in cattle, horses, sheep, goats, and pigs. Scientific protocol available cannot be used for different species, even when washing procedures in combination with trypsin treatment are performed to avoid transmission of diseases, mainly those associated with virus that could be attached in the embryo zona pellucida [18]. Therefore, the susceptibility of the different MOs should be addressed by PDT and further applied to semen and embryo technologies.

14.3 Nanoparticle-Based PDT

Nanotechnology involves the creation of any material, system, or device through the manipulation of matter at very small scale, measuring 1–100 nm. Nanomaterials are defined as small objects that behave as a whole unit regarding to their transport and properties. Moreover, nanomaterials have unique electronic, optical, magnetic, and chemical properties distinct of larger particles of the same material.

Recent developments in nanotechnology allowed improving the use of the PDT for both cancer and infections. In fact, mainly for cancer, the accomplishment of PDT may be partial due to the difficulty in administering PS with low water solubility, which compromises the clinical use of several molecules. Nanotechnology is an interesting approach for PDT mainly because nanoparticles (NPs) (organic and inorganic) can be guided to increase PS concentration at the target and diminish toxic effects to normal tissue and cells. In fact, various types of NP as metallic (silver and gold NP), crystalline (upconversion – rare earth doped), superparamagnetic (superparamagnetic iron oxide nanoparticle, SPION), and semiconductor (quantum dots (QD)) can be functionalized (marked with specific molecules) for use in PDT [19]. Besides, particularly for cancer PDT, NP can accumulate at the tumor site due to increased endocytic activity and leaky vasculature in the tumors. NP can also enhance the solubility of hydrophobic PS.

Incorporation of PS in nanostructured delivery systems, such as polymeric nanoparticles, solid lipid nanoparticles, nanostructured lipid carriers, gold nanoparticles, hydrogels, liposomes, liquid crystals, dendrimers, and cyclodextrin, is an emergent approach to surpass PDT limitations. Thus, the application of nanotechnology offers exciting possibilities to improve cancer and antimicrobial PDT for humans and veterinary medicine.

Different NPs have been used in PDT with distinct interactions between NP and PS. Some examples were presented in Chap. 3. Depending on interaction, NP can be active (NP acts as PS) or passive. Four interactions are described by literature [20] (Fig. 14.1):

1. The PS is embedded in a polymeric NP. In this case, nanoparticles are loaded with PS and are used as carriers to deliver the PS into the target, incorporated on biocompatible and biodegradable matrixes such as liposomes or synthetic and natural polymers (e.g., poly (lactic-co-glycolic acid) (PLGA), chitosan, and cellulose).

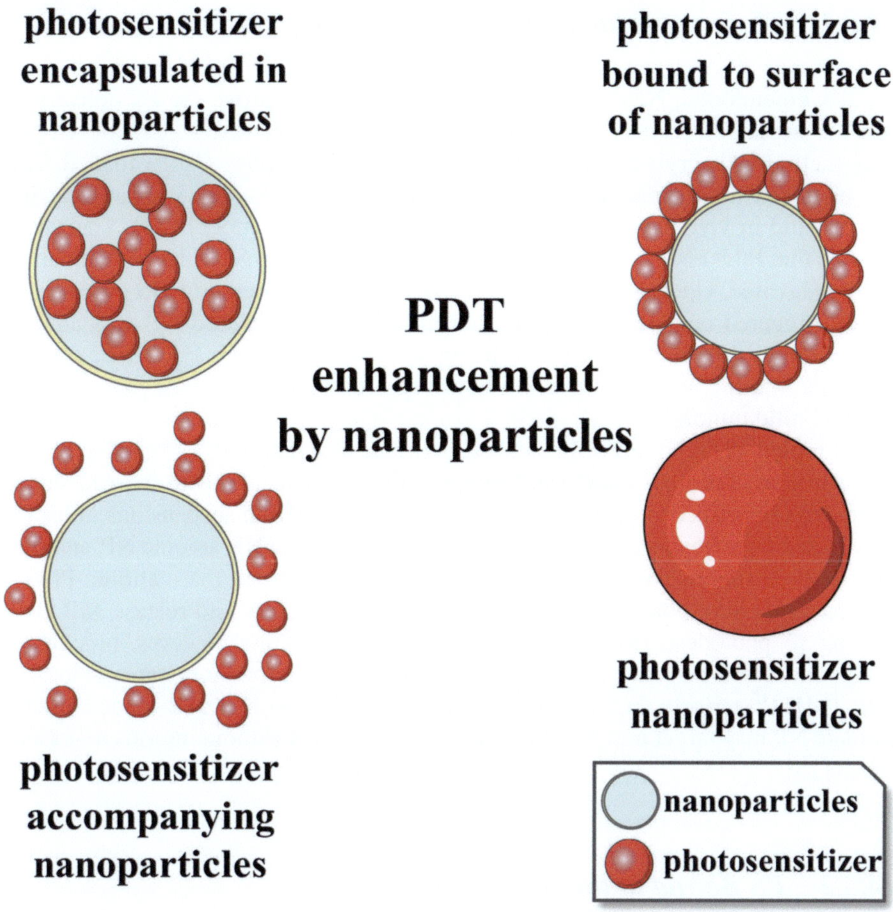

Fig. 14.1 Schematic illustration of the interaction between nanoparticles and photosensitizers to enhance PDT

2. The PS is bound to the NP surface. In this case, the new PS presents better properties compared to original PS.
3. The PS is accompanied by NP. In this case, nanoparticles are used to enhance the photodynamic effect. Metallic NP (gold and silver) and quantum dots have been reported to enhance PDT efficiency in both cancer and antimicrobial PDT [21, 22].
4. The NP acts as the PS. In this case, NP is itself photoactive and able to generate reactive oxygen species (ROS).

Here, we address some works that encompass nanotechnology-based PDT with potential application to veterinary medicine. Our intention is not to provide the peculiarities behind the NP development but to present some studies involving different NPs that were successfully combined to cancer and antimicrobial PDT.

14.3.1 PS Encapsulated in or Immobilized to NP

As abovementioned, NPs can be used as a vehicle to improve the PS delivery at specific sites. For cancer treatment, recently European researchers developed poly-methylmethacrylate core-shell fluorescent nanoparticles (FNP) loaded with the photosensitizer tetrasulfonated aluminum phthalocyanine (Ptl) and carried in vitro and in vivo assays using a human prostate tumor model [23]. Their data showed that Ptl@FNP is internalized by tumor cells and intracellular accumulation of Ptl is favored. Upon irradiation with $\lambda = 680$ nm, they observed ROS production, which triggered cell death. In a murine model, the engineered NP was able to reduce tumor growth with higher efficiency compared to bare Ptl. Thus, authors conclude that the new system could be successfully used to photodynamic treatment of solid tumors.

For topical PDT, NP-based delivery systems are also reported. MB-loaded PLGA nanoparticles of positive charge and with a diameter of about 200 nm showed higher photodynamic effect compared to anionic NP and free MB in suspensions of bacteria isolated from human dental plaque. In biofilms, cationic NP, anionic NP, and free MB showed similar photodynamic action. Authors conclude that cationic PLGA nanoparticles have potential to be used as carriers to diffuse and release MB conducting to photodestruction of bacteria and oral biofilms; however, preclinical assays should be performed to guarantee microbial killing without damage to host cells [24]. Posteriorly, Fontana and collaborators showed that MB-loaded PLGA cationic NP may target oral biofilm safely and fast in rats without injuries to normal tissue [25].

14.3.2 PS Bound to the NP Surface

PS has been bound to the NP surface to prepare new PSs with improved characteristics compared to the former. In the Eshghi's work, authors hypothesized that protoporphyrin IX (PpIX)-conjugated gold NP could improve PS solubility and oxygen singlet quantum yield [26]. PpIX-conjugated gold NP was synthesized, characterized, and used for the delivery of a hydrophobic PS to a cervical cancer cell line. They reported that the PpIX–gold NP conjugate was an excellent carrier for the delivery of surface bound PpIX into HeLa cells. Cellular viability reduction was dependent on conjugate concentration and irradiation time.

Antimicrobial PDT using functionalized NP has also been explored. Tomás et al. reported the functionalization in aqueous media of tiopronin (a thiolate to protect the NP)-gold NPs and ortho-toluidine blue (TBO) to augment the PDT effect on *S. aureus* [27]. TBO was covalently coupled to tiopronin-gold NPs and showed that the minimum bactericidal concentration was at least four times lower than that of free TBO.

14.3.3 PS Alongside NP

When PS accompanies the NP frequently is to enhance the photodynamic action through physical/chemical interactions between PS and NP in the target surroundings. Here, we report localized surface plasmon resonance (LSPR) and Förster (or fluorescence, when both molecules are fluorescent) resonance energy transfer (FRET) to improve PDT.

LSPR is an optical phenomena produced by light when it interacts with noble metal NPs (e.g., gold and silver) that are smaller than the incident wavelength. The electric field of incident light excites the electrons of the conduction band generating localized plasmon oscillations with a resonant frequency that depends on the composition, size, geometry, dielectric environment, and particle-particle separation distance of NPs [28].

Light interaction with PS can be improved by LSPR when LSPR frequency and the PS absorption spectrum overlap. Thus, the field density can be perceived by PS that is placed close to the metallic NP inducing luminescence enhancement to a better excitation of the PS. In fact, gold NP was tested to enhance the antimicrobial effectiveness on *S. aureus* of the PS ortho-toluidine blue (TBO) when irradiated with broad-spectrum light by Narband and collaborators [29]. Bacterial suspension was exposed to white light in the presence of either TBO or a combination of TBO and gold NP (2 nm and 15 nm). Authors observed an increase in bacterial kills concluding that 15 nm gold NPs augment the light-capturing ability of the TBO.

FRET comprises the energy transfer between two photosensitive molecules in close proximity. The donor molecule may transfer energy to the acceptor molecule through nonradiative dipole–dipole coupling (e.g., Coulomb interactions; see Chap. 2), i.e., the donor does not emit a photon that is then absorbed by the acceptor, but instead, the energy is coupled through the molecule dipoles that emit energy in the same manner as a radio antenna [30]. For efficient FRET to occur, the distance and donor and receptor must be too small (<10 nm) since the efficiency of this energy transfer is inversely proportional to the sixth power of the distance between donor and acceptor.

Narband's group also explored FRET to optimize PDT [21]. The authors investigated if CdSe/ZnS quantum dots (QD) (emission maximum at $\lambda = 627$ nm) could enhance the antibacterial activity of TBO (absorption maximum at $\lambda = 630$ nm)-mediated PDT on *S. aureus* and *Streptococcus pyogenes* exposed to white light. Bacterial killing depended on type of MO and TBO/QD ratio. However, authors suggested that enhanced killing seemed to be not attributable to a FRET since QD converted a part of the incident light to the absorption maximum for TBO, which in turn absorbed more light to produce bactericidal radicals.

By the other hand, sulfonated aluminum phthalocyanines (AlPcS) were conjugated with amine-dihydrolipoic acid-coated QD by electrostatic binding [31]. The AlPcS–QD conjugates easily penetrated into human nasopharyngeal carcinoma cells and carried out the FRET in cells, with efficiency around 80 %. Authors used a green laser emitting at $\lambda = 532$ nm, which excited the QD but not the AlPcS, and the cellular AlPcS–QD conjugates damaged most cancer cells via FRET-mediated PDT.

14.3.4 NP as PS

Mostly NP acting as PS are inorganic that strongly absorb ultraviolet (UV) light. However, the use of UV lamps in biomedical sciences could bring safety and health risks. Some strategies to their use as PS encompass the use of sunlight or doped them with other elements (e.g., Er^{3+} and Yb^{3+}) to shift their absorbance toward visible light.

Zinc oxide (ZnO) NP displays an excellent photooxidation activity but with low photocatalytic decomposition. Metal ions (e.g., Ag) can be incorporated to ZnO NP to improve its photocatalytic activity. Thus, Arooj et al. investigated the effects of ZnO/Ag nanocomposites on human malignant melanoma (HT144) and normal (HCEC) cells [32]. The ZnO/Ag nanocomposites killed cancer cells more efficiently than normal cells under daylight exposure. Cytotoxicity was dependent on Ag concentration. Besides the incorporation of Ag into ZnO NP significantly improved their photooxidation capabilities.

Functionalized fullerenes, i.e., fullerenes with attached side chains, are also explored in PDT. The fullerenes are known for their photostability and experience less photobleaching than other PS. As they are insoluble, they can be modified to have a certain degree of lipophilicity. Other modifications can also be carried out to make fullerenes suitable for PDT [33].

Grinholc and colleagues studied the effects in vitro of a C60 fullerene functionalized with one methylpyrrolidinium group (fulleropyrrolidine) on Gram-negative and Gram-positive bacteria, as well as fungal cells under white light exposure. Due to the high antimicrobial activity, authors tested its potential in vivo on *S. aureus*-infected wounds in mice [34]. Fullerene-mediated PDT was efficient to eradicate bacteria, and wounds remained clear up to the third day post-PDT. Incubation of human dermal keratinocytes with fullerene up to 1 μM under illumination did not significantly influence cell viability.

As reported in Chap. 7, light sources used in cancer PDT usually emit between 600 and 700 nm. Thus, cancer PDT still faces some limitation due to poor tissue penetration of these wavelengths compared to near-infrared (NIR) wavelengths (800–1000 nm) to activate PS molecules. Thus, NIR-excited upconversion nanoparticles (UCNPs) emerge as a new strategy that could be used to activate PS molecules in much deeper tissues. UCNPs are usually lanthanide-doped nanocrystals, which emit high-energy photons (e.g., blue light) under excitation by low-energy photons (NIR light). Loading of PS molecules on to UCNP can be by encapsulation, non-covalently physical adsorption, or covalent conjugation [35].

Park et al. were the first to describe effective in vivo PDT through the systemic administration of UCNP-chlorine6 (Ce6) followed by 980-nm irradiation [36]. UCNP-Ce6 was injected in nude mice bearing U87MG tumors through the tail. Accumulated UCNP-Ce6 in tumor was visualized by luminescence and magnetic resonance imaging. Following irradiation, tumor growth of mice was significantly inhibited compared with other control groups. Authors concluded that UCNP-Ce6 presents great potential for multimodal imaging-guided PDT.

Human trials using NP-based PDT are still scarce in literature. In fact, there is one study which reported that MB-loaded poly(lactic-co-glycolic) (PLGA) nanoparticles may be a promising adjunct to treat chronic periodontitis under 660 nm light [37]. However, animal models as reported above demonstrate a promising future to this emerging therapeutic platform.

References

1. Shen J, Chui C, Tao X. Luminous fabric devices for wearable low-level light therapy. Biomed Opt Express. 2013;4(12):2925–37.
2. Lucroy MD. Photodynamic therapy for companion animals with cancer. Vet Clin North Am Small Anim Pract. 2002;32(3):693–702.
3. Truschnegg A, Pichelmayer M, Acham S, Jakse N. Nonsurgical treatment of an epulis by photodynamic therapy. Photodiagnosis Photodyn Ther. 2016;14:1–3.
4. Ribeiro MG, Rodrigues de Farias M, Roesler U, Roth K, Rodigheri SM, Ostrowsky MA, et al. Phenotypic and genotypic characterization of Prototheca zopfii in a dog with enteric signs. Res Vet Sci. 2009;87(3):479–81.
5. Sellera FP, Sabino CP, Ribeiro MS, Gargano RG, Benites NR, Melville PA, et al. In vitro photoinactivation of bovine mastitis related pathogens. Photodiagnosis Photodyn Ther. 2016;13:276–81.
6. Tardivo JP, Adami F, Correa JA, Pinhal MA, Baptista MS. A clinical trial testing the efficacy of PDT in preventing amputation in diabetic patients. Photodiagnosis Photodyn Ther. 2014;11(3):342–50.
7. Tardivo JP, Baptista MS. Treatment of osteomyelitis in the feet of diabetic patients by photodynamic antimicrobial chemotherapy. Photomed Laser Surg. 2009;27(1):145–50.
8. Siqueira EG, Rahal SC, Ribeiro MG, Paes AC, Listoni FP, Vassalo FG. Exogenous bacterial osteomyelitis in 52 dogs: a retrospective study of etiology and in vitro antimicrobial susceptibility profile (2000–2013). Vet Q. 2014;34(4):201–4.
9. Traxler HV, Maguire PJ, Fischetti AJ, Lesser AS. What is your diagnosis? Osteomyelitis of the patella. J Am Vet Med Assoc. 2013;243(11):1529–31.
10. Sherman AB, Clode AB, Gilger BC. Impact of fungal species cultured on outcome in horses with fungal keratitis. Vet Ophthalmol. 2016. doi: 10.1111/vop.12381. [Epub ahead of print].
11. Lahav M, Dueker D, Bhatt PN, Albert DM. Photodynamic inactivation in experimental herpetic keratitis. Arch Ophthalmol. 1975;93(3):207–14.
12. Shih MH, Huang FC. Effects of photodynamic therapy on rapidly growing nontuberculous mycobacteria keratitis. Invest Ophthalmol Vis Sci. 2011;52(1):223–9.
13. Wentink G, Frankena K, Bosch J, Vandehoek J, van den Berg T. Prevention of disease transmission by semen in cattle. Livest Prod Sci. 2000;62(3):207–20.
14. Givens M, Waldrop J. Bovine viral diarrhea virus in embryo and semen production systems. Vet Clin North Am Food Anim Pract. 2004;20(1):21–38.
15. Givens M, Marley M. Pathogens that cause infertility of bulls or transmission via semen. Theriogenology. 2008;70(3):504–7.
16. Sutmoller P, Wrathall A. The risks of disease transmission by embryo transfer in cattle. Rev Sci Technol OIE. 1997;16(1):226–39.
17. Wrathall A, Simmons H, Van Soom A. Evaluation of risks of viral transmission to recipients of bovine embryos arising from fertilisation with virus-infected semen. Theriogenology. 2006;65(2):247–74.
18. Van Soom A, Wrathall A, Herrler A, Nauwynck H. Is the zona pellucida an efficient barrier to viral infection? Reprod Fertil Dev. 2010;22(1):21–31.

19. Jha RK, Jha PK, Chaudhury K, Rana SV, Guha SK. An emerging interface between life science and nanotechnology: present status and prospects of reproductive healthcare aided by nano-biotechnology. Nano Rev. 2014;5:22762. doi: 10.3402/nano.v5.22762.
20. Perni S, Prokopovich P, Pratten J, Parkin IP, Wilson M. Nanoparticles: their potential use in antibacterial photodynamic therapy. Photochem Photobiol Sci. 2011;10(5):712–20.
21. Narband N, Mubarak M, Ready D, Parkin IP, Nair SP, Green MA, et al. Quantum dots as enhancers of the efficacy of bacterial lethal photosensitization. Nanotechnology. 2008;19(44):445102.
22. El-Hussein A, Mfouo-Tynga I, Abdel-Harith M, Abrahamse H. Comparative study between the photodynamic ability of gold and silver nanoparticles in mediating cell death in breast and lung cancer cell lines. J Photochem Photobiol B. 2015;153:67–75.
23. Duchi S, Ramos-Romero S, Dozza B, Guerra-Rebollo M, Cattini L, Ballestri M, et al. Development of near-infrared photoactivable phthalocyanine-loaded nanoparticles to kill tumor cells: an improved tool for photodynamic therapy of solid cancers. Nanomedicine. 2016;12(16):1885–97.
24. Klepac-Ceraj V, Patel N, Song X, Holewa C, Patel C, Kent R, et al. Photodynamic effects of methylene blue-loaded polymeric nanoparticles on dental plaque bacteria. Lasers Surg Med. 2011;43(7):600–6.
25. Fontana CR, Lerman MA, Patel N, Grecco C, Costa CA, Amiji MM, et al. Safety assessment of oral photodynamic therapy in rats. Lasers Med Sci. 2013;28(2):479–86.
26. Eshghi H, Sazgarnia A, Rahimizadeh M, Attaran N, Bakavoli M, Soudmand S. Protoporphyrin IX-gold nanoparticle conjugates as an efficient photosensitizer in cervical cancer therapy. Photodiagnosis Photodyn Ther. 2013;10(3):304–12.
27. Gil-Tomas J, Tubby S, Parkin I, Narband N, Dekker L, Nair S, et al. Lethal photosensitisation of Staphylococcus aureus using a toluidine blue O-tiopronin-gold nanoparticle conjugate. J Mater Chem. 2007;17(35):3739–46.
28. Petryayeva E, Krull U. Localized surface Plasmon resonance: nanostructures, bioassays and biosensing-a review. Anal Chim Acta. 2011;706(1):8–24.
29. Narband N, Tubby S, Parkin I, Gil-Tomas J, Ready D, Nair S, et al. Gold nanoparticles enhance the toluidine blue-induced lethal photosensitisation of staphylococcus aureus. Curr Nanosci. 2008;4(4):409–14.
30. Piston D, Kremers G. Fluorescent protein FRET: the good, the bad and the ugly. Trends Biochem Sci. 2007;32(9):407–14.
31. Li L, Zhao JF, Won N, Jin H, Kim S, Chen JY. Quantum dot-aluminum phthalocyanine conjugates perform photodynamic reactions to kill cancer cells via fluorescence resonance energy transfer. Nanoscale Res Lett. 2012;7(1):386.
32. Arooj S, Nazir S, Nadhman A, Ahmad N, Muhammad B, Ahmad I, et al. Novel ZnO:Ag nanocomposites induce significant oxidative stress in human fibroblast malignant melanoma (Ht144) cells. Beilstein J Nanotechnol. 2015;6:570–82.
33. Huang YY, Sharma SK, Yin R, Agrawal T, Chiang LY, Hamblin MR. Functionalized fullerenes in photodynamic therapy. J Biomed Nanotechnol. 2014;10(9):1918–36.
34. Grinholc M, Nakonieczna J, Fila G, Taraszkiewicz A, Kawiak A, Szewczyk G, et al. Antimicrobial photodynamic therapy with fulleropyrrolidine: photoinactivation mechanism of Staphylococcus aureus, in vitro and in vivo studies. Appl Microbiol Biotechnol. 2015;99(9):4031–43.
35. Wang C, Cheng L, Liu Z. Upconversion nanoparticles for photodynamic therapy and other cancer therapeutics. Theranostics. 2013;3(5):317–30.
36. Park YI, Kim HM, Kim JH, Moon KC, Yoo B, Lee KT, et al. Theranostic probe based on lanthanide-doped nanoparticles for simultaneous in vivo dual-modal imaging and photodynamic therapy. Adv Mater. 2012;24(42):5755–61.
37. de Freitas LM, Calixto GM, Chorilli M, Giusti JS, Bagnato VS, Soukos NS, et al. Polymeric nanoparticle-based photodynamic therapy for chronic periodontitis in vivo. Int J Mol Sci. 2016;17(5):769. doi: 10.3390/ijms17050769.

Index

F.P. Sellera et al. (eds.), *Photodynamic Therapy in Veterinary Medicine: From Basics to Clinical Practice*, DOI 10.1007/978-3-319-45007-0

MIX
Papier aus verantwortungsvollen Quellen
Paper from responsible sources
FSC® C105338

If you have any concerns about our products,
you can contact us on
ProductSafety@springernature.com

In case Publisher is established outside the EU,
the EU authorized representative is:
Springer Nature Customer Service Center GmbH
Europaplatz 3, 69115 Heidelberg, Germany

Printed by Libri Plureos GmbH
in Hamburg, Germany